INTERSCIENCE TRACTS IN PURE AND APPLIED MATHEMATICS

1.

D. Montgomery and L. Zippin
TOPOLOGICAL TRANSFORMATION GROUPS

2.

Fritz John
PLANE WAVES AND SPHERICAL MEANS
Applied to Partial Differential Equations

3.

E. Artin
GEOMETRIC ALGEBRA

Additional volumes in preparation

INTERSCIENCE TRACTS
IN PURE AND APPLIED MATHEMATICS

Number 3

GEOMETRIC ALGEBRA

By E. Artin

INTERSCIENCE PUBLISHERS, INC., NEW YORK

a division of John Wiley & Sons, Inc., New York • London • Sydney

GEOMETRIC ALGEBRA

E. ARTIN
Princeton University, Princeton, New Jersey

Wiley Classics Library Edition Published 1988

WILEY

INTERSCIENCE PUBLISHERS, INC., NEW YORK

a division of John Wiley & Sons, Inc., New York · London · Sydney

10

LIBRARY OF CONGRESS CATALOG CARD NUMBER 57-6109

ISBN 0 470 03432 7

ISBN 0-471-60839-4 (pbk.).

PRINTED IN THE UNITED STATES OF AMERICA

TO NATASCHA

PREFACE

Many parts of classical geometry have developed into great independent theories. Linear algebra, topology, differential and algebraic geometry are the indispensable tools of the mathematician of our time. It is frequently desirable to devise a course of geometric nature which is distinct from these great lines of thought and which can be presented to beginning graduate students or even to advanced undergraduates. The present book has grown out of lecture notes for a course of this nature given at New York University in 1955. This course centered around the foundations of affine geometry, the geometry of quadratic forms and the structure of the general linear group. I felt it necessary to enlarge the content of these notes by including projective and symplectic geometry and also the structure of the symplectic and orthogonal groups. Lack of space forced me to exclude unitary geometry and the quadratic forms of characteristic 2.

I have to thank in the first place my wife who helped greatly with the preparation of the manuscript and with the proofs. My thanks go also to George Bachman who with the help of Bernard Sohmer wrote the notes for the original course, to Larkin Joyner who drew the figures, and to Susan Hahn for helping with the proofreading.

E. Artin

SUGGESTIONS FOR THE USE OF THIS BOOK

The most important point to keep in mind is the fact that Chapter I should be used mainly as a reference chapter for the proofs of certain isolated algebraic theorems. These proofs have been collected so as not to interrupt the main line of thought in later chapters.

An inexperienced reader should start right away with Chapter II. He will be able to understand for quite a while, provided he knows the definition of a field and the rudiments of group theory. More knowledge will be required from §8 on and he may then work his way through the first three paragraphs and the beginning of §9 of Chapter I. This will enable him to read the rest of Chapter II except for a few harder algebraic theorems which he should skip in a first reading.

This skipping is another important point. It should be done whenever a proof seems too hard or whenever a theorem or a whole paragraph does not appeal to the reader. In most cases he will be able to go on and later on he may return to the parts which were skipped.

The rest of the book definitely presupposes a good knowledge of §4 of Chapter I[1]. The content of this paragraph is of such a fundamental importance for most of modern mathematics that every effort should be devoted to its mastery. In order to facilitate this, the content of §4 is illustrated in §5 by the theory of linear equations and §6 suggests an exercise on which the reader can test his understanding of the preceding paragraphs. If he can do this exercise then he should be well equipped for the remainder of the book.

Chapter III gives the theory of quadratic and of skew symmetric bilinear forms in a geometric language. For a first reading the symplectic geometry may be disregarded.

Chapter IV is almost independent of the preceding chapters. If the reader does not find the going too heavy he may start the book with Chapter IV. But §4 of Chapter I will be needed.

Chapter V connects, so to speak, the ideas of Chapters III and

[1]It is sufficient to know it for finite dimensional spaces only.

and IV. The problems of Chapter IV are investigated for the groups introduced in Chapter III.

Any one of these chapters contains too much material for an advanced undergraduate course or seminar. I could make the following suggestions for the content of such courses.

1) The easier parts of Chapter II.

2) The linear algebra of the first five paragraphs of Chapter I followed by selected topics from Chapter III, either on orthogonal or on symplectic geometry.

3) The fundamental theorem of projective geometry, followed by some parts of Chapter IV.

4) Chapter III, but with the following modification:

All that is needed from §4 of Chapter I is the statement:

If W^* is the space orthogonal to a subspace W of a non-singular space V then $\dim W + \dim W^* = \dim V$. This statement could be obtained from the naive theory of linear equations and the instructor could supply a proof of it. Our statement implies then $W^{**} = W$ and no further reference to §4 of Chapter I is needed.

CONTENTS

Chapter IV

The General Linear Group

Chapter V

The Structure of Symplectic and Orthogonal Groups

CHAPTER 1

Preliminary Notions

1. Notions of set theory

We begin with a list of the customary symbols:

$a \in S$ means a is an element of the set S.
$S \subset T$ means S is a subset of T.
$S \cap T$ means the intersection of the sets S and T; should it be empty we call the sets disjoint.
$S \cup T$ stands for the union of S and T.

$\bigcap_i S_i$ and $\bigcup_i S_i$ stand for intersection and union of a family of indexed sets. Should S_i and S_j be disjoint for $i \neq j$ we call $\bigcup_i S_i$ a disjoint union of sets. Sets are sometimes defined by a symbol $\{\cdots\}$ where the elements are enumerated between the parenthesis or by a symbol $\{x|A\}$ where A is a property required of x; this symbol is read: "the set of all x with the property A". Thus, for example:

$$S \cap T = \{x|x \in S, x \in T\}.$$

If f is a map of a non-empty set S into a set T, i.e., a function $f(s)$ defined for all elements $s \in S$ with values in T, then we write either

$$f : S \to T \quad \text{or} \quad S \xrightarrow{f} T.$$

If $S \xrightarrow{f} T$ and $T \xrightarrow{g} U$ we also write $S \xrightarrow{f} T \xrightarrow{g} U$. If $s \in S$ then we can form $g(f(s)) \in U$ and thus obtain a map from S to U denoted by $S \xrightarrow{gf} U$. Notice that the associative law holds trivially for these "products" of maps. The order of the two factors gf comes from the notation $f(s)$ for the image of the elements. Had we written $(s)f$ instead of $f(s)$, it would have been natural to write fg instead of gf. Although we will stick (with rare exceptions) to the notation $f(s)$ the reader should be able to do everything in the reversed notation. Sometimes it is even convenient to write s^f instead of $f(s)$ and we should notice that in this notation $(s^f)^g = s^{gf}$.

If $S \xrightarrow{f} T$ and $S_0 \subset S$ then the set of all images of elements of S_0 is denoted by $f(S_0)$; it is called the image of S_0. This can be done particularly for S itself. Then $f(S) \subset T$; should $f(S) = T$ we call the map *onto* and say that f maps S onto T.

Let T_0 be a subset of T. The set of all $s \in S$ for which $f(s) \in T_0$ is called the inverse image of T_0 and is denoted by $f^{-1}(T_0)$. Notice that $f^{-1}(T_0)$ may very well be empty, even if T_0 is not empty. Remember also that f^{-1} is *not* a map. By $f^{-1}(t)$ for a certain $t \in T$ we mean the inverse image of the set $\{t\}$ with the one element t. It may happen that $f^{-1}(t)$ never contains more than one element. Then we say that f is a one-to-one *into* map. If f is onto and one-to-one into, then we say that f is one-to-one onto, or a "*one-to-one correspondence.*" In this case only can f^{-1} be interpreted as a map $T \xrightarrow{f^{-1}} S$ and is also one-to-one onto. Notice that $f^{-1}f : S \to S$ and $ff^{-1} : T \to T$ and that both maps are identity maps on S respectively T.

If $t_1 \neq t_2$ are elements of T, then the sets $f^{-1}(t_1)$ and $f^{-1}(t_2)$ are disjoint. If s is a given element of S and $f(s) = t$, then s will be in $f^{-1}(t)$, which shows that S is the disjoint union of all the sets $f^{-1}(t)$:

$$S = \bigcup_{t \in T} f^{-1}(t).$$

Some of the sets $f^{-1}(t)$ may be empty. Keep only the non-empty ones and call S_f the set whose elements are these non-empty sets $f^{-1}(t)$. Notice that the elements of S_f are *sets* and not elements of S. S_f is called a quotient set and its elements are also called equivalence classes. Thus, s_1 and s_2 are in the same equivalence class if and only if $f(s_1) = f(s_2)$. Any given element s lies in precisely one equivalence class; if $f(s) = t$, then the equivalence class of s is $f^{-1}(t)$.

We construct now a map $f_1 : S \to S_f$ by mapping each $s \in S$ onto its equivalence class. Thus, if $f(s) = t$, then $f_1(s) = f^{-1}(t)$. This map is an onto map.

Next we construct a map $f_2 : S_f \to f(S)$ by mapping the non-empty equivalence class $f^{-1}(t)$ onto the element $t \in f(S)$. If $t \in f(S)$, hence $t = f(s)$, then t is the image of the equivalence class $f^{-1}(t)$ and of no other. This map f_2 is therefore one-to-one and onto. If $s \in S$ and $f(s) = t$, then $f_1(s) = f^{-1}(t)$ and the image of $f^{-1}(t)$ under the map f_2 is t. Therefore, $f_2f_1(s) = t$.

Finally we construct a very trivial map $f_3 : f(S) \to T$ by setting $f_3(t) = t$ for $t \in f(S)$. This map should not be called identity since it is a map of a subset into a possibly bigger set T. A map of this kind is called an injection and is of course one-to-one into. For $f(s) = t$ we had $f_2f_1(s) = t$ and thus $f_3f_2f_1(s) = t$. We have $S \xrightarrow{f_1} S_f \xrightarrow{f_2} f(S) \xrightarrow{f_3} T$, so that $f_3f_2f_1 : S \rightarrow T$. We see that our original map f is factored into three maps

$$f = f_3f_2f_1 \, .$$

To repeat: f_1 is onto, f_2 is a one-to-one correspondence and f_3 is one-to-one into. We will call this the canonical factoring of the map f. The word "*canonical*," or also "*natural*," is applied in a rather loose sense to any mathematical construction which is unique in as much as no free choices of objects are used in it.

As an example, let G and H be groups, and $f : G \rightarrow H$ a homomorphism of G into H, i.e., a map for which $f(xy) = f(x)f(y)$ holds for all $x, y \in G$. Setting $x = y = 1$ (unit of G) we obtain $f(1) = 1$ (unit in H). Putting $y = x^{-1}$, we obtain next $f(x^{-1}) = (f(x))^{-1}$. We will now describe the canonical factoring of f and must to this effect first find the quotient set G_f . The elements x and y are in the same equivalence class if and only if $f(x) = f(y)$ or $f(xy^{-1}) = 1$ or also $f(y^{-1}x) = 1$; denoting by K the inverse image of 1 this means that both $xy^{-1} \in K$ and $y^{-1}x \in K$ (or $x \in Ky$ and $x \in yK$). The two cosets yK and Ky are therefore the same and the elements x which are equivalent to y form the coset yK. If we take y already in K, hence y in the equivalence class of 1 we obtain $yK = K$, so that K is a group. The equality of left and right cosets implies that K is an invariant subgroup and our quotient set merely the factor group G/K. The map f_1 associates with each $x \in G$ the coset xK as image: $f_1(x) = xK$. The point now is that f_1 is a homomorphism (onto). Indeed $f_1(xy) = xyK = xyK \cdot K = x \cdot Ky \cdot K = xK \cdot yK = f_1(x)f_1(y)$.

This map is called the *canonical* homomorphism of a group onto its factor group.

The map f_2 maps xK onto $f(x) : f_2(xK) = f(x)$. Since $f_2(xK \cdot yK) = f_2(xy \cdot K) = f(xy) = f(x)f(y) = f_2(xK)f_2(yK)$ it is a homomorphism. Since it is a one-to-one correspondence it is an isomorphism and yields the statement that the factor group G/K is isomorphic to the image group $f(G)$. The invariant subgroup K of G is called the kernel of the map f.

The map f_3 is just an injection and therefore an isomorphism into H.

2. *Theorems on vector spaces*

We shall assume that the reader is familiar with the notion and the most elementary properties of a vector space but shall repeat its definition and discuss some aspects with which he may not have come into contact.

DEFINITION 1.1. A right vector space V over a field k (k need not be a commutative field) is an additive group together with a composition Aa of an element $A \in V$ and an element $a \in k$ such that $Aa \in V$ and such that the following rules hold:

1) $(A + B)a = Aa + Ba$, 2) $A(a + b) = Aa + Ab$,

3) $(Aa)b = A(ab)$, 4) $A \cdot 1 = A$,

where $A, B \in V$, $a, b \in k$ and where 1 is the unit element of k.

In case of a left vector space the composition is written aA and similar laws are supposed to hold.

Let V be a right vector space over k and S an arbitrary subset of V. By a linear combination of elements of S one means a finite sum $A_1a_1 + A_2a_2 + \cdots + A_ra_r$ of elements A_i of S. It is easy to see that the set $\langle S \rangle$ of all linear combinations of elements of S forms a subspace of V and that $\langle S \rangle$ is the smallest subspace of V which contains S. If S is the empty set we mean by $\langle S \rangle$ the smallest subspace of V which contains S and, since 0 is in any subspace, the space $\langle S \rangle$ consists of the zero vector alone. This subspace is also denoted by 0.

We call $\langle S \rangle$ the space generated (or spanned) by S and say that S is a system of generators of $\langle S \rangle$.

A subset S is called independent if a linear combination $A_1a_1 + A_2a_2 + \cdots + A_ra_r$ of distinct elements of S is the zero vector only in the case when all $a_i = 0$. The empty set is therefore independent.

If S is independent and $\langle S \rangle = V$ then S is called a basis of V. This means that every vector of V is a linear combination of distinct elements of S and that such an expression is unique up to trivial terms $A \cdot 0$.

If T is independent and L is any system of generators of V then T can be "completed" to a basis of V by elements of L. This means that there exists a subset L_0 of L which is disjoint from T such that the set $T \cup L_0$ is a basis of V. The reader certainly knows this statement, at least when V is finite dimensional. The proof for the infinite dimensional case necessitates a transfinite axiom such as Zorn's lemma but a reader who is not familiar with it may restrict all the following considerations to the finite dimensional case.

If V has as basis a finite set S, then the number n of elements of S ($n = 0$ if S is empty) depends only on V and is called the dimension of V. We write $n = \dim V$. This number n is then the maximal number of independent elements of V and any independent set T with n elements is a basis of V. If U is a subspace of V, then $\dim U \leq \dim V$ and the equal sign holds only for $U = V$.

The fact that V does not have such a finite basis is denoted by writing $\dim V = \infty$. A proper subspace U of V may then still have the dimension ∞. (One could introduce a more refined definition of $\dim V$, namely the cardinality of a basis. We shall not use it, however, and warn the reader that certain statements we are going to make would not be true with this refined definition of dimension.)

The simplest example of an n-dimensional space is the set of all n-tuples of elements of k with the following definitions for sum and product:

$$(x_1, x_2, \cdots, x_n) + (y_1, y_2, \cdots, y_n) = (x_1 + y_1, \cdots, x_n + y_n),$$

$$(x_1, x_2, \cdots, x_n)a = (x_1a, x_2a, \cdots, x_na).$$

If U and W are subspaces of V (an arbitrary space), then the space spanned by $U \cup W$ is denoted by $U + W$. Since a linear combination of elements of U is again an element of U we see that $U + W$ consists of all vectors of the form $A + B$ where $A \in U$ and $B \in W$. The two spaces U and W may be of such a nature that an element $U + W$ is *uniquely* expressed in the form $A + B$ with $A \in U$, $B \in W$. One sees that this is the case if and only if $U \cap W = 0$. We say then that the sum $U + W$ is *direct* and use the symbol $U \oplus W$. Thus one can write $U \oplus W$ for $U + W$ if and only if $U \cap W = 0$.

If U_1, U_2, U_3 are subspaces and if we can write $(U_1 \oplus U_2) \oplus U_3$, then an expression $A_1 + A_2 + A_3$ with $A_i \in U_i$ is unique and thus one can also write $U_1 \oplus (U_2 \oplus U_3)$. We may therefore leave out the parenthesis: $U_1 \oplus U_2 \oplus U_3$. An intersection of subspaces is always a subspace.

Let U now be a subspace of V. We remember that V was an additive group. This allows us to consider the additive factor group V/U

whose elements are the cosets $A + U$. ($A + U$ for an arbitrary but *fixed* $A \in V$ means the set of all vectors of the form $A + B$, $B \in U$.) Equality $A_1 + U = A_2 + U$ of two cosets means $A_1 - A_2 \in U$, addition is explained by $(A_1 + U) + (A_2 + U) = (A_1 + A_2) + U$. We also have the canonical map

$$\varphi : V \to V/U$$

which maps $A \in V$ onto the coset $A + U$ containing A. The map φ is an additive homomorphism of V onto V/U. We make V/U into a vector space by defining the composition of an element $A + U$ of V/U and an element $a \in k$ by:

$$(A + U)\cdot a = Aa + U.$$

One has first to show that this composition is *well* defined, i.e., does not depend on the particular element A of the coset $A + U$. But if $A + U = B + U$, then $A - B \in U$, hence $(A - B)a \in U$ which shows $Aa + U = Ba + U$. That the formal laws of Definition 1.1 are satisfied is pretty obvious. For the canonical map φ we have

$$\varphi(Aa) = Aa + U = (A + U)\cdot a = \varphi(A)\cdot a$$

in addition to the fact that φ is an additive homomorphism. This suggests

DEFINITION 1.2. Let V and W be two right vector spaces (W not necessarily a subspace of V) over k. A map $f : V \to W$ is called a homomorphism of V into W if

1) $f(A + B) = f(A) + f(B)$, $\quad A \in V$ and $B \in V$,

2) $f(Aa) = f(A)\cdot a$, $\quad A \in V$ and $a \in k$.

Should f be a one-to-one correspondence, we call f an isomorphism of V onto W and we denote the mere existence of such an isomorphism by $V \simeq W$ (read: "V isomorphic to W").

Notice that such a homomorphism is certainly a homomorphism of the additive group. The notion of kernel U of f is therefore already defined, $U = f^{-1}(0)$, the set of all $A \in V$ for which $f(A) = 0$. If $A \in U$ then $f(Aa) = f(A)\cdot a = 0$ so that $Aa \in U$. This shows that U is not only a subgroup but even a subspace of V.

Let U be an arbitrary subspace of V and $\varphi : V \to V/U$ the canonical map. Then it is clear that φ is a homomorphism of V onto V/U.

The zero element of V/U is the image of 0, hence U itself. The kernel consists of all $A \in V$ for which

$$\varphi(A) = A + U = U.$$

It is therefore the given subspace U. One should mention the special case $U = 0$. Each coset $A + U$ is now the set with the single element A and may be identified with A. Strictly speaking we have only a canonical isomorphism $V/0 \simeq V$ but we shall write $V/0 = V$.

Let us return to any homomorphism $f : V \to W$ and let U be the kernel of V. Since f is a homomorphism of the additive groups we have already the canonical splitting

$$V \xrightarrow{f_1} V/U \xrightarrow{f_2} f(V) \xrightarrow{f_3} W$$

where $f_1(A) = A + U$ is the canonical map $V \to V/U$, where $f_2(A + U) = f(A)$ and, therefore,

$$f_2((A + U)a) = f_2(Aa + U) = f(Aa) = f(A)a = f_2(A + U)a$$

and where f_3 is the injection. All three maps are consequently homomorphisms between the vector spaces, and f_2 is an isomorphism onto. We have, therefore,

THEOREM 1.1. *To a given homomorphism $f : V \to W$ with kernel U we can construct a canonical isomorphism f_2 mapping V/U onto the image space $f(V)$.*

Suppose now that U and W are given subspaces of V. Let φ be the canonical map $V \xrightarrow{\varphi} V/U$. The restriction ψ of φ to the given subspace W is a canonically constructed homomorphism $W \xrightarrow{\psi} V/U$. What is $\psi(W)$? It consists of all cosets $A + U$ with $A \in W$. The union of these cosets forms the space $W + U$, the cosets $A + U$ are, therefore, the stratification of $W + U$ by cosets of the subspace U of $W + U$. This shows $\psi(W) = (U + W)/U$. What is the kernel of ψ? For all elements $A \in W$ we have $\psi(A) = \varphi(A)$. But φ has, in V, the kernel U so that ψ has $U \cap W$ as kernel. To ψ we can construct the canonical map ψ_2 which exhibits the isomorphism of $W/(U \cap W)$ with the image $(U + W)/U$. Since everything was canonical we have

THEOREM 1.2. *If U and W are subspaces of V then $(U + W)/U$ and $W/(U \cap W)$ are canonically isomorphic.*

In the special case $V = U \oplus W$ we find that V/U and $W/(U \cap W) = W/0 = W$ are canonically isomorphic. Suppose now that only the subspace U of V is given. Does there exist a subspace W such that $V = U \oplus W$? Such a subspace shall be called supplementary to U. Let S be a basis of U and complete S to a basis $S \cup T$ of V where S and T are disjoint. Put $W = \langle T \rangle$, then $U + W = V$ and obviously $V = U \oplus W$. This construction involves choices and is far from being canonical.

THEOREM 1.3. *To every subspace U of V one can find (in a non-canonical way) supplementary spaces W for which $V = U \oplus W$. Each of these supplementary subspaces W is, however, canonically isomorphic to the space V/U. If $V = U \oplus W_1 = U \oplus W_2$ then W_1 is canonically isomorphic to W_2.*

If $f : V \to W$ is an isomorphism *into* then the image $f(S)$ of a basis of V will at least be independent. One concludes the inequality $\dim V \leq \dim W$. Should f be also onto then equality holds.

In our construction of W we also saw that $\dim V = \dim U + \dim W$ and since $W \simeq V/U$ one obtains

$$\dim V = \dim U + \dim V/U$$

hence also, whenever $V = U \oplus W$, that

$$\dim V = \dim U + \dim W.$$

Let now $U_1 \subset U_2 \subset U_3$ be subspaces of V. Find subspaces W_2 and W_3 such that

$$U_2 = U_1 \oplus W_2 , \qquad U_3 = U_2 \oplus W_3$$

and, therefore,

$$U_3 = U_1 \oplus (W_2 \oplus W_3).$$

We have $\dim U_2/U_1 = \dim W_2$, $\dim U_3/U_2 = \dim W_3$ and $\dim U_3/U_1 = \dim(W_2 \oplus W_3) = \dim W_2 + \dim W_3$. Thus we have proved: if $U_1 \subset U_2 \subset U_3$, then

$$\dim U_3/U_1 = \dim U_2/U_1 + \dim U_3/U_2 . \tag{1.1}$$

Let now U and W be two given subspaces of V. Use (1.1) for $U_1 = 0$, $U_2 = U$, $U_3 = U + W$. We obtain

$$\begin{aligned}\dim(U + W) &= \dim U + \dim(U + W)/U \\ &= \dim U + \dim W/(U \cap W).\end{aligned}$$

If we add on both sides $\dim(U \cap W)$ and use $\dim W/(U \cap W) + \dim(U \cap W) = \dim W$ we get

$$\dim(U + W) + \dim(U \cap W) = \dim U + \dim W.$$

Next we use (1.1) for $U_1 = U \cap W$, $U_2 = W$, $U_3 = V$:

$$\begin{aligned}\dim V/(U \cap W) &= \dim W/(U \cap W) + \dim V/W \\ &= \dim (U + W)/U + \dim V/W.\end{aligned}$$

If we add $\dim V/(U + W)$ and use

$$\dim V/(U + W) + \dim (U + W)/U = \dim V/U$$

we obtain

$$\dim V/(U + W) + \dim V/(U \cap W) = \dim V/U + \dim V/W.$$

If the dimension of V is finite all subspaces of V have finite dimension. If, however, $\dim V = \infty$, then our interest will be concentrated on two types of subspaces U. Those whose dimension is finite and, on the other hand, those which are extremely large, namely those which have a finite dimensional supplement. For spaces of the second type $\dim U = \infty$ but $\dim V/U$ is finite; $\dim U$ tells us very little about U, but $\dim V/U$ gives us the amount by which U differs from the whole space V. We give, therefore, to $\dim V/U$ a formal status by

Definition 1.3. The dimension of the space V/U is called the codimension of U:

$$\operatorname{codim} U = \dim V/U.$$

The various results we have obtained are expressed in

Theorem 1.4. *The following rules hold between dimensions and codimensions of subspaces:*

$$\dim U + \operatorname{codim} U = \dim V, \tag{1.2}$$

$$\dim(U + W) + \dim(U \cap W) = \dim U + \dim W, \tag{1.3}$$

$$\operatorname{codim}(U + W) + \operatorname{codim}(U \cap W) = \operatorname{codim} U + \operatorname{codim} W. \tag{1.4}$$

These rules are of little value unless the terms on one side are finite (then those on the other side are also) since an ∞ could not be transposed to the other side by subtraction.

Spaces of dimension one are called lines, of dimension two planes and spaces of codimension one are called hyperplanes.

3. *More detailed structure of homomorphisms*

Let V and V' be right vector spaces over a field k and denote by $\text{Hom}(V, V')$ the set of all homomorphisms of V into V'. We shall make $\text{Hom}(V, V')$ into an abelian additive group by defining an addition:

If f and g are $\varepsilon \text{Hom}(V, V')$, let $f + g$ be the map which sends the vector $X \varepsilon V$ onto the vector $f(X) + g(X)$ of V'; in other words,

$$(f + g)(X) = f(X) + g(X).$$

That $f + g$ is a homomorphism and that the addition is associative and commutative is easily checked. The map which sends every vector $X \varepsilon V$ onto the 0 vector of V' is obviously the 0 element of $\text{Hom}(V, V')$ and shall also be denoted by 0. If $f \varepsilon \text{Hom}(V, V')$, then the map $-f$ which sends X onto $-(f(X))$ is a homomorphism and indeed $f + (-f) = 0$. The group property is established.

In special situations it is possible to give more structure to $\text{Hom}(V, V')$ and we are going to investigate some of the possibilities.

a) $V' = V$.

An element of $\text{Hom}(V, V)$ maps V into V; one also calls it an endomorphism of V. If $f, g \varepsilon \text{Hom}(V, V)$, then it is possible to combine them to a map $gf : V \xrightarrow{f} V \xrightarrow{g} V$ as we did in §1 : $gf(X) = g(f(X))$. One sees immediately that gf is also a homomorphism of $V \to V$.

Since

$$(g_1 + g_2)f(X) = g_1f(X) + g_2f(X) = (g_1f + g_2f)(X)$$

and

$$g(f_1 + f_2)(X) = g(f_1(X) + f_2(X)) = gf_1(X) + gf_2(X)$$
$$= (gf_1 + gf_2)X,$$

we see that both distributive laws hold; $\text{Hom}(V, V)$ now becomes a ring. This ring has a unit element, namely the identity map.

The maps f which are a one-to-one correspondence lead to an inverse map f^{-1} which is also in $\text{Hom}(V, V)$. These maps f form therefore a group under multiplication. All of Chapter IV is devoted to the study of this group if dim V is finite.

Let us now investigate some elementary properties of Hom(V, V) if dim $V = n$ is finite. Let $f \in$ Hom(V, V) and let U be the kernel of f. Then $V/U \simeq f(V)$ so that the dimension of the image $f(V)$ is $n -$ dim U. This shows that f is an onto map if and only if dim $U = 0$, i.e., if and only if f is an isomorphism into.

Let $A_1, A_2, \cdots, A_n$ be a basis of V and set $f(A_i) = B_i$. If $X = A_1x_1 + A_2x_2 + \cdots + A_nx_n \in V$ then

$$f(X) = B_1x_1 + B_2x_2 + \cdots + B_nx_n . \tag{1.5}$$

Conversely choose any n vectors $B_i \in V$ and define a map f by (1.5). One sees easily that $f \in$ Hom(V, V) and that $f(A_i) = B_i$. Consequently f is completely determined by the images B_i of the basis elements A_i and the B_i can be any system of n vectors of V. If we express each B_i by the basis A_ν,

$$f(A_j) = B_j = \sum_{\nu=1}^{n} A_\nu a_{\nu j}, \qquad j = 1, 2, \cdots, n,$$

then we see that f is described by an n-by-n matrix (a_{ij}) where i is the index of the rows and j the index of the columns.

Let $g \in$ Hom(V, V) be given by the matrix (b_{ij}) which means that

$$g(A_j) = \sum_{\nu=1}^{n} A_\nu b_{\nu j} .$$

Then

$$(f + g)(A_j) = \sum_{\nu=1}^{n} A_\nu(a_{\nu j} + b_{\nu j})$$

and

$$(fg)(A_j) = f\Big(\sum_{\nu=1}^{n} A_\nu b_{\nu j}\Big) = \sum_{\nu=1}^{n} f(A_\nu)b_{\nu j}$$

$$= \sum_{\nu=1}^{n}\Big(\sum_{\mu=1}^{n} A_\mu a_{\mu\nu}\Big)b_{\nu j}$$

$$= \sum_{\mu=1}^{n} A_\mu\Big(\sum_{\nu=1}^{n} a_{\mu\nu}b_{\nu j}\Big).$$

We see that $f + g$ is described by the matrix $(a_{ij} + b_{ij})$ and fg by $(\sum_{\nu=1}^{n} a_{i\nu}b_{\nu j})$. This is the reason for defining addition and multiplication of matrices by

$$(a_{ij}) + (b_{ij}) = (a_{ij} + b_{ij}),$$

$$(a_{ij})\cdot(b_{ij}) = \left(\sum_{\nu=1}^{n} a_{i\nu}b_{\nu j}\right).$$

Under this definition of addition and multiplication the correspondence $f \to (a_{ij})$ becomes an isomorphism between $\mathrm{Hom}(V, V)$ and the ring of all n-by-n matrices.

This isomorphism is far from canonical since it depends on the choice of the basis A_i for V.

Let g be another element of $\mathrm{Hom}(V, V)$, but suppose that g is one-to-one. Let (b_{ij}) be the matrix associated with the element gfg^{-1} of $\mathrm{Hom}(V, V)$. The meaning of the matrix (b_{ij}) is that

$$gfg^{-1}(A_j) = \sum_{\nu=1}^{n} A_\nu b_{\nu j}\,.$$

If we apply g^{-1} to this equation it becomes

$$f(g^{-1}(A_j)) = \sum_{\nu=1}^{n} g^{-1}(A_\nu)\cdot b_{\nu j}\,.$$

Since g^{-1} is any one-to-one onto map of V the vectors $g^{-1}(A_\nu)$ are another basis of V, and g can be chosen in such a way that $g^{-1}(A_\nu)$ is any given basis of V. Looking at the equation from this point of view we see that the matrix (b_{ij}) is the one which would describe f if we had chosen $g^{-1}(A_\nu)$ as basis of V. Therefore:

The matrix describing f in terms of the new basis is the same as the one describing gfg^{-1} in terms of the old basis A_ν. In this statement g was the map which carries the "new basis" $g^{-1}(A_\nu)$ into the old one,

$$g(g^{-1}(A_\nu)) = A_\nu\,.$$

This g is, therefore, a fixed map once the new basis is given. Suppose now that $f \to A$, $g \to D$ are the descriptions of f and g in terms of the original basis. Then $gfg^{-1} \to DAD^{-1}$. The attitude should be that g is fixed, determined by the old and the new basis, and that f ranges over $\mathrm{Hom}(V, V)$. We can state

THEOREM 1.5. *The ring* Hom(V, V) *is isomorphic to the ring of all n-by-n matrices with elements in k. The isomorphism depends on the choice of a basis. Let g be an element of* Hom(V, V) *which carries a selected new basis into the old one and suppose that $g \rightarrow D$ describes g in terms of the old basis. If $f \rightarrow A$ is the description of any f in terms of the old basis, then DAD^{-1} is the description of this same f in terms of the new basis.*

Mathematical education is still suffering from the enthusiams which the discovery of this isomorphism has aroused. The result has been that geometry was eliminated and replaced by computations. Instead of the intuitive maps of a space preserving addition and multiplication by scalars (these maps have an immediate geometric meaning), matrices have been introduced. From the innumerable absurdities—from a pedagogical point of view—let me point out one example and contrast it with the direct description.

Matrix method: A product of a matrix A and a vector X (which is then an n-tuple of numbers) is defined; it is also a vector. Now the poor student has to swallow the following definition:

A vector X is called an eigen vector if a number λ exists such that

$$AX = \lambda X.$$

Going through the formalism, the characteristic equation, one then ends up with theorems like: If a matrix A has n distinct eigen values, then a matrix D can be found such that DAD^{-1} is a diagonal matrix.

The student will of course learn all this since he will fail the course if he does not.

Instead one should argue like this: Given a linear transformation f of the space V into itself. Does there exist a line which is kept fixed by f? In order to include the eigen value 0 one should then modify the question by asking whether a line is mapped *into* itself. This means of course for a vector spanning the line that

$$f(X) = \lambda X.$$

Having thus motivated the problem, the matrix A describing f will enter only for a moment for the actual computation of λ. It should disappear again. Then one proves all the customary theorems without ever talking of matrices and asks the question: Suppose we can find a basis of V which consists of eigen vectors; what does this imply for the geometric description of f? Well, the space is stretched in the various directions of the basis by factors which are the eigen values. Only then does one ask what this means for the description of f by a matrix in terms of this basis. We have obviously the diagonal form.

I should of course soften my reproach since books have appeared lately which stress this point of view so that improvements are to be expected.

It is my experience that proofs involving matrices can be shortened by 50% if one throws the matrices out. Sometimes it can not be done; a determinant may have to be computed.

Talking of determinants we assume that the reader is familiar with them. In Chapter IV we give a definition which works even in the non-commutative case; let us right now stick to the commutative case. If k is a commutative field and $f \in \mathrm{Hom}(V, V)$, then we define the determinant of f as follows: Let A be a matrix describing f and put $\det f = \det A$.

If we use a new basis, A has to be replaced by DAD^{-1}, the determinant becomes $\det D \cdot \det A \cdot (\det D)^{-1}$ by the multiplication theorem of determinants; since k is commutative $\det D$ cancels and we see that the map

$$f \to \det f = \det A$$

is well defined and canonical. If g corresponds to the matrix B, then fg corresponds to the matrix AB and the multiplication theorem shows $\det fg = \det f \cdot \det g$.

THEOREM 1.6. *There exists a well defined map*

$$\mathrm{Hom}(V, V) \to k \qquad (\textit{if } k \textit{ is commutative})$$

called the determinant of an endomorphism f. It satisfies

$$\det (fg) = \det f \cdot \det g.$$

In view of this fact it should be possible to describe $\det f$ in an intrinsic manner. The reader will find such a description in Bourbaki, Algèbre, Chapter III.

b) V' is a two-sided space.

DEFINITION 1.4. Suppose that V' is both a right and a left vector space over k and that we have the additional rule $(aA)b =$

$a(Ab)$ for all $a, b \in k$ and $A \in V'$. Then we call V' a two-sided space over k.

THEOREM 1.7. *If V is a right and V' a two-sided vector space over k, then* Hom(V, V') *can be made into a left vector space over k in a canonical way.*

To this effect we have to define a product af for $a \in k$ and $f \in \text{Hom}(V, V')$. We mean by af the function which sends the vector $X \in V$ onto the vector $a \cdot f(X)$ in V'. Since

$$(af)(X + Y) = a \cdot f(X + Y) = a(f(X) + f(Y))$$
$$= a \cdot f(X) + a \cdot f(Y) = (af)(X) + (af)(Y)$$

and

$$(af)(Xb) = a \cdot f(Xb) = a \cdot (f(X)b) = (af(X))b = ((af)(X)) \cdot b,$$
$$af \in \text{Hom}(V, V').$$

The equations

$$(a(f + g))(X) = a((f + g)(X)) = a(f(X) + g(X))$$
$$= (af)(X) + (ag)(X) = (af + ag)(X),$$
$$((a + b)f)(X) = (a + b) \cdot f(X) = af(X) + bf(X) = (af + bf)(X),$$
$$(1 \cdot f)(X) = 1 \cdot f(X) = f(X)$$

show that Hom(V, V') is a left space over k.

The question arises: Can one make any right space V' in a natural way into a two-sided space by defining a product aX from the left? One thinks of course of the definition $aX = Xa$. But then $a(bX) = (bX)a = (Xb)a = X(ba) = (ba)X$, whereas we should have obtained $(ab)X$. This "natural" definition works only if the field k is commutative.

THEOREM 1.7′. *If V and V' are vector spaces over a commutative field k, then* Hom(V, V') *can be made in a natural way into a vector space over k.*

If k is again any field, then the most important example of a two-sided vector space V' is the field k itself if one defines addition and multiplication as they are defined in the field. We shall investigate this case in the next paragraph.

4. *Duality and pairings*

DEFINITION 1.5. If V is a right vector space over k, then the set $\hat{V} = \mathrm{Hom}(V, k)$ is a left vector space over k called the dual of V. The elements $\varphi, \psi, \cdots$ of $\hat{V}$ are called functionals of V. To repeat the definition of a functional φ: it is a map $V \xrightarrow{\varphi} k$ such that

$$1)\quad \varphi(A + B) = \varphi(A) + \varphi(B), \qquad 2)\quad \varphi(Aa) = \varphi(A)\cdot a.$$

The operations between functionals are

$$3)\quad (\varphi + \psi)(A) = \varphi(A) + \psi(A), \qquad 4)\quad (a\varphi)(A) = a\cdot\varphi(A).$$

If V is a left vector space over k we also define the dual space $\hat{V} = \mathrm{Hom}(V, k)$. In order to obtain complete symmetry we write, however, $(A)\varphi$ instead of $\varphi(A)$. $\hat{V}$ is a right vector space over k.

The notation will become still simpler if we write φA instead of $\varphi(A)$ (and $A\varphi$ instead of $(A)\varphi$ if V is a left space). Rewriting 1) and 3) these rules take on the form of "distributive laws" and 2) and 4) become "associative laws". In other words let us regard φA as a kind of "product" of an element $\varphi \in \hat{V}$ and an element $A \in V$ such that $\varphi A \in k$.

This change of notation suggests a generalisation where a right space V and a left space W are given (W not necessarily the dual of V), together with a product $AB \in k$ for $A \in W$, $B \in V$.

DEFINITION 1.6. If W is a left and V a right vector space over k, we say that a pairing of W and V into F is given, provided a product $AB \in k$ is defined for all $A \in W$ and all $B \in V$ such that the following rules hold:

$$1)\quad A(B_1 + B_2) = AB_1 + AB_2\,, \qquad 2)\quad A(Bb) = (AB)b,$$

$$3)\quad (A_1 + A_2)B = A_1B + A_2B, \qquad 4)\quad (aA)B = a(AB).$$

We notice that $\hat{V}$ and V are naturally paired into k.

Our task is to study $\hat{V}$ more closely and to investigate general pairings.

Let $\{A_i\}$ be a basis of the right space V where the indices i range over some set I which we have used for indexing. Let $\varphi \in \hat{V}$, and put $\varphi A_i = a_i \in k$. An element $X \in V$ can be written uniquely in the form $X = \sum_i A_i x_i$ where $x_i \neq 0$ holds only for a *finite* number of indices, since X must be a *finite* linear combination of $\{A_i\}$. Then

$$\varphi X = \sum_i \varphi A_i x_i = \sum_i a_i x_i .$$

This shows that φ is known if all the a_i are known. Select conversely an $a_i \in k$ for each index i, not restricted by any condition and define a function $\varphi(X)$ by

$$\varphi(X) = \sum_i a_i x_i .$$

The sum on the right side makes sense since $x_i \neq 0$ holds only for a finite number of the subscripts i. The two equations $\varphi(X + Y) = \varphi(X) + \varphi(Y)$ and $\varphi(Xb) = \varphi(X)b$ are immediately checked. Since $A_i = \sum_\nu A_\nu \delta_{\nu i}$, where, as usual, $\delta_{ii} = 1$ and $\delta_{ji} = 0$ for $j \neq i$, we get also $\varphi(A_i) = \sum_\nu a_\nu \delta_{\nu i} = a_i$. Thus we have

THEOREM 1.8. *If $\{A_i\}$ is a basis of V, then for randomly given $a_i \in k$ (one for each i) there is one and only one $\varphi \in \hat{V}$ such that $\varphi A_i = a_i$.*

Let i be one of the subscripts. Denote by φ_i the functional for which $\varphi_i A_i = 1$ and $\varphi_i A_j = 0$ for $i \neq j$. In other words $\varphi_i A_j = \delta_{ij}$.

A finite linear combination of the φ_i has the form $\varphi = \sum_i b_i \varphi_i$ where $b_i \neq 0$ holds only for a finite number of i. The φ which we obtain in this way form the subspace W_0 of $\hat{V}$ which is spanned by the φ_i. For such a φ we get

$$\varphi A_j = \sum_i b_i \varphi_i A_j = \sum_i b_i \delta_{ij} = b_j .$$

This means that φ alone already determines the b_j as values of φ on the basis vectors. The linear combination $\sum_i b_i \varphi_i$ is, therefore, unique, the φ_i are independent and consequently a basis of W_0. Since there are as many φ_i as there are A_i, we get $\dim W_0 = \dim V$. W_0 can be described as the set of those functionals φ of V for which $\varphi A_i \neq 0$ holds only for a finite number of i. Should $\dim V$ be finite, then W_0 contains obviously all functionals of V, i.e., $W_0 = \hat{V}$ and we get $\dim V = \dim \hat{V}$. If $\dim V = \infty$, then $\dim W_0 = \infty$ and, since $W_0 \subset \hat{V}$ and our notion of dimension is very crude, we also have $\dim \hat{V} = \infty$. W_0 is then certainly not the whole space $\hat{V}$. If $A = \sum_i A_i a_i \in V$ and $A \neq 0$, then at least one $a_j \neq 0$. For this j we get $\varphi_j A \neq 0$. By the definition of a functional we know trivially that only the zero functional vanishes on all of V. Now we see an analogue: If $A \in V$, and if $\varphi A = 0$ for all $\varphi \in \hat{V}$, then $A = 0$.

Let us now state our results.

THEOREM 1.9. *We always have* dim $\hat{V}$ = dim V. *If we know* $\varphi A = 0$ *for all* A, *then* $\varphi = 0$; *if we know* $\varphi A = 0$ *for all* φ, *then* $A = 0$. *Let* dim $V = n$ *be finite. To a given basis* $\{A_i\}$ *of* V *we can find a "dual basis"* $\{\varphi_i\}$ *of* $\hat{V}$ *where* $\varphi_i A_j = \delta_{ij}$.

Turning our attention to pairings we suppose that a pairing of the left space W and the right space V into k is given.

DEFINITION 1.7. If $A \in W$, $B \in V$ and $AB = 0$ we shall say that A is orthogonal to B. If W_0 is a subspace of W and V_0 a subspace of V, we say that W_0 is orthogonal to V_0 provided $AB = 0$ for $A \in W_0$ and $B \in V_0$. If V_0 is a subspace of V, one sees easily that the set V_0^* of all vectors in W which are orthogonal to V_0 is a *subspace* of W: $V_0^* \subset W$. Similarly each given subspace W_0 of W gives rise to a subspace W_0^* of V. We have trivially $V_0 \subset (V_0^*)^*$. Abbreviating we write V_0^{**} instead of $(V_0^*)^*$. Of special importance is the subspace V^* of W consisting of all vectors of W which are orthogonal to all vectors of V. We shall call V^* the left kernel of our pairing. Similarly we call the subspace W^* of V the right kernel of our pairing.

Notice that Theorem 1.9 tells us that in the pairing of $\hat{V}$ and V both kernels are 0.

Suppose that in our pairing of W and V into k the *left* kernel is 0. Each $A \in W$ gives us a function φ_A on V with values in k, namely

$$\varphi_A(X) = AX.$$

The function φ_A is easily seen to be a homomorphism $V \to k$, i.e., $\varphi_A \in \hat{V}$.

This suggests studying the map $W \to \hat{V}$ which maps the vector A onto φ_A. We have

$$\varphi_{A+B}X = (A + B)X = AX + BX = (\varphi_A + \varphi_B)X$$

and

$$\varphi_{aA}X = (aA)X = a(AX) = a\varphi_A X.$$

The two equations $\varphi_{A+B} = \varphi_A + \varphi_B$ and $\varphi_{aA} = a\varphi_A$ mean that our map $W \to \hat{V}$ is a homomorphism. The kernel of this map consists of those $A \in W$ for which φ_A is the zero function: $AX = 0$ for all $X \in V$. However we had assumed that the left kernel of our pairing is zero. Hence $A = 0$. Our map $W \to \hat{V}$ is, therefore, an isomorphism into (not always onto).

Similarly, if the right kernel is 0, one obtains an isomorphism into: $V \to \hat{W}$.

Suppose again that the *left* kernel is 0. Let W_0 be a subspace of W and W_0^* the subspace of V which is orthogonal to W_0. We can find in a natural way a new pairing of the space W_0 and the space V/W_0^* into k by defining as product of a vector $X \varepsilon W_0$ and of a coset $Y + W_0^*$ in V/W_0^* the element XY of k:

$$X \cdot (Y + W_0^*) = XY.$$

This pairing is well defined. Indeed $Y + W_0^* = Y_1 + W_0^*$ implies $Y - Y_1 \varepsilon W_0^*$ and since X is in W_0 we have $X(Y - Y_1) = 0$ or $XY = XY_1$. That this new multiplication satisfies the axioms of a pairing is obvious. What is the *right* kernel? $Y + W_0^*$ will be in the right kernel if $XY = 0$ for all $X \varepsilon W_0$. This means $Y \varepsilon W_0^*$ and, therefore, $Y + W_0^* = W_0^*$, the zero element of the space V/W_0^*. The right kernel of our new pairing is 0. We can use the previously established method to construct an isomorphism into (canonical):

$$V/W_0^* \to \hat{W}_0 .$$

What is this map? Given an element $A + W_0^*$ of V/W_0^*; the functional of W_0 associated with it is

$$X \cdot (A + W_0^*) = XA \qquad (X \text{ ranging over } W_0).$$

If V_0 is a given subspace of V, we can also define a natural pairing of V_0^* and V/V_0 by setting

$$X \cdot (Y + V_0) = XY, \qquad X \varepsilon V_0^*, \qquad Y + V_0 \varepsilon V/V_0 .$$

As before this pairing is well defined and satisfies the axioms. This time we ask for the left kernel; X will lie in the left kernel if $XY = 0$ for all $Y \varepsilon V$. But since we have assumed that the left kernel of our original pairing is 0, this means that the left kernel of our new pairing is 0. We obtain, therefore, an isomorphism into:

$$V_0^* \to \widehat{V/V_0} .$$

The inexperienced reader is urged not to give up but to go over all definitions and mappings again and again until he sees everything in full clarity. We formulate our results:

THEOREM 1.10. *Let W and V be paired into k and assume that* *the left kernel V^* of our pairing is 0. Let W_0 be a subspace of W and V_0 a subspace of V. There exist natural isomorphisms into:*

$$(1.6)\qquad V/W_0^* \to \hat{W}_0\,,$$

$$(1.7)\qquad V_0^* \to \widehat{V/V_0}\,.$$

Since these maps are isomorphisms into, the dimensions of the spaces on the left are $\leq$ to the dimensions of the spaces on the right. Under the symbol dim the ^ on top of a space can be dropped because of Theorem 1.9. We obtain $\dim V/W_0^* \leq \dim W_0$ and $\dim V_0^* \leq \dim V/V_0$, inequalities which we can also write in the form

$$\operatorname{codim} W_0^* \leq \dim W_0 \quad \text{and} \quad \dim V_0^* \leq \operatorname{codim} V_0\,.$$

If we put $V_0 = W_0^*$ in the second inequality and combine it with the first, we get

$$\dim W_0^{**} \leq \operatorname{codim} W_0^* \leq \dim W_0\,.$$

And now we remember the trivial fact that $W_0 \subset W_0^{**}$ so that $\dim W_0 \leq \dim W_0^{**}$. Therefore we get equality

$$(1.8)\qquad \dim W_0^{**} = \operatorname{codim} W_0^* = \dim W_0\,.$$

The main significance of this formula appears in the case when $\dim W_0$ is finite. Since $W_0 \subset W_0^{**}$, we see that simply $W_0^{**} = W_0$. In the isomorphism (1.6) both spaces have the same finite dimension, our map is, therefore, onto and thus V/W_0^* may be regarded naturally as the dual of W_0 . The map (1.7) should be used for $V_0 = W_0^*$ and becomes an isomorphism of W_0 into $\widehat{V/W_0^*}$. This map is onto again and we see now that each of the spaces W_0 and V/W_0^* is naturally the dual of the other.

In (1.8) no restriction on W_0 is necessary; we can use the formula on $W_0 = W$. We obtain

$$(1.9)\qquad \operatorname{codim} W^* = \dim W.$$

We should mention that our results are true for *all* subspaces of W if $\dim W$ itself is finite.

What can we do if the left kernel V^* is not 0? We make a new pairing (by now you are used to this procedure) between W/V^* and V by defining

$$(X + V^*)Y = XY, \qquad X + V^* \in W/V^*, \qquad Y \in V.$$

The element $X + V^*$ lies in the left kernel if $XY = 0$ for all $Y \in V$. This means $X \in V^*$, and hence $X + V^* = V^*$, the zero element of V/V^*. The left kernel is now zero. The right kernel is obviously the old W^*. Equation (1.9) tells us that the codimension of W^* in V is the same as the dimension of our left factor which is W/V^*. We obtain, therefore,

$$\dim V/W^* = \dim W/V^*. \tag{1.10}$$

Suppose now that both kernels V^* and W^* are zero and that dim W is finite. (1.10) shows that dim V is also finite and equals dim W. In this case we can use all our results unrestrictedly on subspaces of both W and V. If we choose for instance $W_0 = W$, then $W_0^* = 0$ and $V/W_0^* = V$; we see that each of the spaces W and V is naturally the dual of the other. The reader should again visualize the map: A in W corresponds to the functional $AX = \varphi_A X$ and $A \to \varphi_A$ is one-to-one onto. Let us still have a look at the correspondence $W_0 \leftrightarrow W_0^*$ of a subspace $W_0 \subset W$ and the subspace $W_0^* \subset V$. Any subspace $V_0 \subset V$ is obtainable from a W_0; we have merely to put $W_0 = V_0^*$. And distinct subspaces of W give distinct images since $W_0^* = W_1^*$ implies (by starring) $W_0 = W_1$. It is therefore a one-to-one correspondence which is lattice inverting, i.e., an inclusion $W_0 \subset W_1$ implies $W_0^* \supset W_1^*$ (a strict inclusion becomes strict again). Again let us collect our results.

THEOREM 1.11. *Assume W and V paired into k:*

a) $\dim W/V^* = \dim V/W^*$; *in particular, if one of the spaces W/V^* and V/W^* is finite dimensional, the other one is also, and the dimensions are equal.*

b) *If the left kernel is 0 and $W_0 \subset W$, then*

$$\dim W_0 = \operatorname{codim} W_0^* = \dim W_0^{**}. \tag{1.11}$$

If $\dim W_0$ is finite, then $W_0^{**} = W_0$ and each of the spaces W_0 and V/W_0^* is naturally the dual of the other.

c) *If both kernels are 0 and $\dim W$ is finite, then each of the spaces W and V is naturally the dual of the other. The correspondence $W_0 \leftrightarrow W_0^*$ is one-to-one between the subspaces of W and the subspaces of V. It reverses any inclusion relation (in the strict sense).*

In case our pairing is the one between $\hat{V}$ and V we can strengthen the results. We know already that both kernels are 0. Let V_0 be any subspace of V (not necessarily of finite dimension) and consider the map (1.7) of Theorem 1.10 : $V_0^* \to \widehat{V/V_0}$. It is an isomorphism into and we shall show that it is onto. Let φ be any functional of V/V_0 . We construct a function $\bar{\varphi}$ on the space V by defining

$$\bar{\varphi}(X) = \varphi(X + V_0).$$

Then

$$\bar{\varphi}(X + Y) = \varphi(X + Y + V_0) = \varphi((X + V_0) + (Y + V_0))$$
$$= \varphi(X + V_0) + \varphi(Y + V_0) = \bar{\varphi}(X) + \bar{\varphi}(Y)$$

and

$$\bar{\varphi}(Xa) = \varphi(Xa + V_0) = \varphi((X + V_0)a) = \varphi(X + V_0)a = \bar{\varphi}(X)a.$$

This implies that $\bar{\varphi}$ is a functional of V. Suppose $X \in V_0$, then $\bar{\varphi}(X) = \varphi(X + V_0) = \varphi(V_0) = \varphi(\text{zero element of } V/V_0) = 0.$ $\bar{\varphi}$ vanishes on all of V_0 and, therefore, belongs to V_0^* . We contend that $\bar{\varphi}$ has the given φ as image under the isomorphism (1.7) and this will show the ontoness of our map. What is its image? We had to make the new pairing of V_0^* and V/V_0 by defining $Y \cdot (X + V_0) = YX$ ($Y \in V_0^*, X \in V$); for $Y = \bar{\varphi} \in V_0^*$ this means $\bar{\varphi} \cdot (X + V_0) = \bar{\varphi}X$ and this function $\bar{\varphi} \cdot (X + V_0)$ on V/V_0 is the image. But $\bar{\varphi}X$ was by definition $\varphi(X + V_0)$ so that this function on V/V_0 is indeed φ.

The space V_0^* is now naturally the dual of V/V_0 . Let $A \notin V_0$. The coset $A + V_0$ is not the zero element of V/V_0 so that a functional $\varphi \in \widehat{V/V_0}$ can be found such that $\varphi(A + V_0) \neq 0$ (Theorem 1.9). The corresponding $\bar{\varphi} \in V_0^*$ gives then $\bar{\varphi}(A) = \varphi(A + V_0) \neq 0$. This vector A is, therefore, not orthogonal to all of V_0^* and we conclude that a vector orthogonal to all of V_0^* must, by necessity, lie in V_0 .

Since V_0 is trivially orthogonal to V_0^* we see that we have also proved $V_0^{**} = V_0$. Finally, we set in formula (1.11) of Theorem 1.11: $\dim W_0 = \operatorname{codim} W_0^*$, for W_0 the space V_0^* . Since $W_0^* = V_0$, we obtain $\dim V_0^* = \operatorname{codim} V_0$ which supplements the analogue of (1.11), $\dim V_0 = \operatorname{codim} V_0^*$, which is true in all pairings. Thus we see in our special case that there is a one-to-one correspondence $W_0 \leftrightarrow W_0^*$ between subspaces $W_0 \subset W$ of *finite* dimension and subspaces of V with finite codimension. Indeed, if $\operatorname{codim} V_0$ is finite, then $\dim V_0^* = \operatorname{codim} V_0$ is finite and $W_0 = V_0^*$ is the space which gives $W_0^* = V_0$; there is only one such space. If W_0 is given and of finite dimension, then (1.11) shows codim $W_0^* = \dim W_0$ and $W_0 = V_0^*$ gives $V_0 = W_0^*$.

THEOREM 1.12. *Consider the pairing between* $\hat{V} = W$ *and* V. *If* V_0 *is any subspace of* V, *then* $V_0^{**} = V_0$ *and* V_0^* *is naturally the dual of* V/V_0. *We have not only* $\dim V_0 = \operatorname{codim} V_0^*$ *(established in Theorem* 1.11), *but also* $\dim V_0^* = \operatorname{codim} V_0$. *The correspondence* $W_0 \leftrightarrow W_0^*$ *is one-to-one between all subspaces* $W_0 \subset W$ *of finite dimension and all subspaces of* V *with finite codimension. Similar results would not hold in general for subspaces of* V.

Let us look especially at a hyperplane V_0 of V. Then codim $V_0 = 1$, hence dim $V_0^* = 1$. Let $V_0^* = \langle \varphi \rangle$ ($\varphi \neq 0$). Since $V_0^{**} = V_0$ we see that the vectors X of V_0 can be characterized as the solutions of the equation $\varphi X = 0$. Any $\psi \in \hat{V}$ such that $\psi X = 0$ for all $X \in V_0$ must lie in V_0^* and is therefore a left multiple of φ. If we start with any $\varphi \neq 0$ and put $W_0 = \langle \varphi \rangle$ then dim $W_0 = 1$ and hence codim $W_0^* = 1$. The solutions of $\varphi X = 0$ form a hyperplane.

The proof of these simple facts about hyperplanes is burdened by too much theory. Let us see whether we can not get them from scratch:

Take a functional $\varphi \neq 0$ of V. Map $V \to k$ by sending $X \to \varphi X$. This is a homomorphism by definition of a functional. Since $\varphi \neq 0$ there is some non-zero image b and since k (as right k space) is 1-dimensional, the map is onto. Let V_0 be the kernel. Then $V/V_0 \simeq k$, whence dim V/V_0 = codim $V_0 = 1$. Start conversely with a hyperplane V_0. Consider the canonical map $V \to V/V_0$ with kernel V_0. V/V_0 is 1-dimensional, hence isomorphic (not canonical) to k. The map $V \to V/V_0 \to k$ is then a functional φ with kernel V_0. Take any vector $A \notin V_0$ and two functionals φ, ψ with kernel V_0. Then $\varphi(A) = a \neq 0$, $\psi(A) = b \neq 0$ and the functional $\varphi - ab^{-1}\psi$ will map V_0 and A onto 0. It vanishes on V and is therefore $0 : \varphi = ab^{-1}\psi$.

THEOREM 1.13. *A hyperplane* V_0 *of* V *can be described as the set of solutions* $\varphi X = 0$ *where* $\varphi \neq 0$ *is an element of* $\hat{V}$ *and conversely any* $\varphi \neq 0$ *of* $\hat{V}$ *gives in this way a hyperplane.* φ *is determined by* V_0 *up to a left factor* $\neq 0$ *of* k.

5. *Linear equations*

We illustrate the theory developed thus far by applying it to the theory of linear equations. The beginner should not expect that we will be able to develop miracle methods for solving equations. An actual solution is still best found by the elementary method of successive elimination.

Let

$$
(1.12)\qquad
\begin{aligned}
a_{11}x_1 + a_{12}x_2 + \cdots + a_{1n}x_n &= b_1 ,\\
a_{21}x_1 + a_{22}x_2 + \cdots + a_{2n}x_n &= b_2 ,\\
\cdots\cdots\cdots\cdots\cdots\cdots\cdots\cdots&\cdots\cdots\\
a_{m1}x_1 + a_{m2}x_2 + \cdots + a_{mn}x_n &= b_m ,
\end{aligned}
$$

be a system of m equations with n unknown x_i whose coefficients a_{ij} and right sides b_i are given elements of a field k.

The following objects will play a role in the discussion:

1) The matrix (a_{ij}) of the coefficients with m rows and n columns.

2) A *right* n-dimensional vector space V over k and a fixed basis $E_1 , E_2 , \cdots , E_n$ of V. A solution $(x_1 , x_2 , \cdots , x_n)$ of (1.12) shall be interpreted in V by the "solution vector" $X = E_1x_1 + E_2x_2 + \cdots + E_nx_n$.

3) The dual space $\hat{V}$ with the basis $\varphi_1 , \varphi_2 , \cdots , \varphi_n$ dual to $E_1 , E_2 , \cdots , E_n$. With the i-th row of (a_{ij}) we associate the functional

$$\psi_i = a_{i1}\varphi_1 + a_{i2}\varphi_2 + \cdots + a_{in}\varphi_n \qquad (i = 1, 2, \cdots , m).$$

If $X = E_1x_1 + E_2x_2 + \cdots + E_nx_n$ is any vector of V then

$$\psi_iX = a_{i1}x_1 + a_{i2}x_2 + \cdots + a_{in}x_n ;$$

(this follows easily from $\varphi_iE_j = \delta_{ij}$).

4) The subspace W of $\hat{V}$ spanned by $\psi_1 , \psi_2 , \cdots , \psi_m$. Its dimension r is the maximal number of linearly independent vectors among the ψ_i and consequently the maximal number of left linearly independent row vectors of (a_{ij}). We call r therefore the left row rank of (a_{ij}).

5) The m-dimensional *right* space S_m of m-tuples of elements of k. In it we place the n column vectors $A_1 , A_2 , \cdots , A_n$ of the matrix (a_{ij}) and also the vector $B = (b_1 , b_2 , \cdots , b_m)$. The vectors $A_1 , A_2 , \cdots , A_n$ will span a subspace U of S_m and its dimension shall be called the right column rank of the matrix (a_{ij}).

In our setup the equations (1.12) can be rewritten in two ways. In the space S_m they are obviously equivalent with the vector equation

$$(1.13) \qquad A_1x_1 + A_2x_2 + \cdots + A_nx_n = B.$$

But they can also be written as $\psi_i X = b_i$, or as

$$(1.14) \qquad B = (\psi_1 X, \psi_2 X, \cdots, \psi_m X), \qquad X \varepsilon V.$$

Our equations need not have a solution. The problem of solvability is stated in the following way: Consider the matrix (a_{ij}) as given and fixed. For which vectors B does a solution exist?

Equation (1.13) tells us that the vectors B for which a solution exists must lie in the subspace U of S_m.

Equation (1.14) tells us that B must lie in the image space of the following map of $V \xrightarrow{f} S_m$:

$$f(X) = (\psi_1 X, \psi_2 X, \cdots, \psi_m X), \qquad X \varepsilon V.$$

This map f is obviously a homomorphism. Its kernel consists of those vectors X for which $\psi_i X = 0$ $(i = 1, \cdots, m)$, hence of the vectors X which are orthogonal to $\psi_1, \psi_2, \cdots, \psi_m$ and, therefore, orthogonal to the space $W \subset \hat{V}$ which they span. The kernel is therefore W^*. We see that the image space $U \simeq V/W^*$, hence $\dim U = \operatorname{codim} W^*$. But $\operatorname{codim} W^* = \dim W$, and we obtain $\dim U = \dim W$. We have therefore the rule:

$$\text{left row rank of } (a_{ij}) = \text{right column rank of } (a_{ij}),$$

a rule which facilitates sometimes the computation of ranks.

In most applications the answer we have found thus far is satisfactory: The equations have a solution if and only if B belongs to a certain r-dimensional subspace U of S_m, the point being that we know now that r is the *left* row rank of (a_{ij}).

Frequently one asks when the equations have a solution for *all* $B \varepsilon S_m$. This means $U = S_m$ or $r = m$, and is true if and only if the rows of (a_{ij}) are left linearly independent.

Suppose now that B is of such a nature that the equations have a solution and let X_0 be a special solution: $\psi_i X_0 = b_i$ $(i = 1, 2, \cdots, m)$. Then the equations can be rewritten as

$$\psi_i(X - X_0) = 0$$

and mean that $X - X_0$ must belong to the subspace W^* of V. The solutions consist of the coset $X_0 + W^*$. We have, therefore, the familiar rule: general solution = special solution + general solution of the homogeneous equations.

When is the solution unique? This means $W^* = 0$ and consequently $W = \hat{V}$; therefore, $r = n$. Let us review the extreme cases for r:

$$r = m \qquad \text{means solvability for all } B;$$

$$r = n \qquad \text{means uniqueness of the solution, if it exists.}$$

Should $m = n$ (the frequent case of as many equations as there are unknowns), then $r = n$ means solvability for all B as well as uniqueness of the solution, and is also equivalent with $W^* = 0$ which implies that the homogeneous equations have only the trivial solution. This fact is used frequently in applications.

The description $X_0 + W^*$ which we have given for the solutions calls for a geometric language.

Let V_0 be any subspace of V. A coset $A + V_0$ shall be called a linear variety. If the coset $A + V_0$ is merely given as a set of vectors, then one can get back the subspace V_0 by subtracting from each vector of $A + V_0$ a special one, say A. The subspace V_0 shall be called the direction of the linear variety. (The intuitive picture is, of course, to think of vectors as "arrows" from the origin and to think of the linear variety $A + V_0$ as consisting of the endpoints of every "arrow" in $A + V_0$.) We can now say that the solutions of our equations form a linear variety which passes through a special solution and has the direction W^*. If $A + V_0$ is a linear variety, then dim V_0 shall be called the dimension of the linear variety. For the solutions of our equations the dimension is dim $W^* =$ codim W and, therefore, $n - r$.

One may also ask for the converse. Given a linear variety $A + V_0$. What can be said about all equations $\psi X = b$ ($\psi \in \hat{V}$) which are satisfied by all $X \in A + V_0$? They must be satisfied by A, which gives $b = \psi(A)$, and thus they have the form $\psi(X - A) = 0$ or $\psi(V_0) = 0$; ψ must, therefore, belong to V_0^*. Any $\psi \in V_0^*$ gives really an equation: set $b = \psi A$; since $\psi(X - A) = 0$, if $X \in A + V_0$, we get $\psi(X) = b$. If we put $r =$ dim V_0^* and let $\psi_1, \psi_2, \cdots, \psi_r$ be a basis of V_0^*, then $\psi_i X = \psi_i A$ ($i = 1, \cdots, r$) is a set of linear equations and their solution satisfies $\psi_i(X - A) = 0$ ($i = 1, \cdots, r$) so that $X - A \in V_0^{**} = V_0$; $A + V_0$ is the set of solutions, and dim $V_0 =$ codim $V_0^* = n - r$.

The elementary geometry of linear varieties is very simple. Let

$L_1 = A + V_1$ and $L_2 = B + V_2$ be two linear varieties with directions V_1 and V_2. They need not intersect. When is $L_1 \cap L_2$ not empty? If and only if a vector $X_1 \varepsilon V_1$ and a vector $X_2 \varepsilon V_2$ exist such that $A + X_1 = B + X_2$ or $A - B = (-X_1) + X_2$. This means $A - B \varepsilon V_1 + V_2$. If $L_1 \cap L_2$ is not empty and C a common vector, then we can write $L_1 = C + V_1$, $L_2 = C + V_2$, hence $L_1 \cap L_2 = C + (V_1 \cap V_2)$. The direction of $L_1 \cap L_2$ is, therefore, $V_1 \cap V_2$. Its dimension (still for non-empty $L_1 \cap L_2$) is that of $V_1 \cap V_2$. We must also consider the "join" $L_1 \circ L_2$ of L_1 and L_2, the smallest linear variety containing both L_1 and L_2. Its direction must contain $A - B$ and all differences of V_1 and V_2, hence all vectors of $\langle A - B\rangle + V_1 + V_2$. The join contains B, hence certainly $B + \langle A - B\rangle + V_1 + V_2$. But this is a linear variety and contains L_1 and L_2, hence

$$L_1 \circ L_2 = B + \langle A - B\rangle + V_1 + V_2 .$$

Its direction is $\langle A - B\rangle + V_1 + V_2$. We notice that there is a case distinction:

1) $L_1 \cap L_2$ is not empty, then $A - B \varepsilon V_1 + V_2$ and

$$L_1 \circ L_2 = B + V_1 + V_2 , \quad \dim(L_1 \circ L_2) = \dim(V_1 + V_2).$$

We obtain

$$\dim(L_1 \cap L_2) + \dim(L_1 \circ L_2) = \dim L_1 + \dim L_2 .$$

2) $L_1 \cap L_2$ is empty. Then $A - B \not\varepsilon V_1 + V_2$ and

$$\dim(L_1 \circ L_2) = 1 + \dim(V_1 + V_2).$$

Let us illustrate this in a few special cases. If $\dim L_1 = \dim L_2 = 0$, then $L_1 = A$, $L_2 = B$, $V_1 = V_2 = 0$; if $L_1 \neq L_2$, then they do not meet and

$$\dim(L_1 \circ L_2) = 1;$$

$L_1 \circ L_2$ is the unique "line" through A and B.

Suppose $n = \dim V = 2$, $\dim L_1 = \dim L_2 = 1$. If $L_1 \cap L_2$ is empty we have $\dim(L_1 \circ L_2) = 1 + \dim(V_1 + V_2) \leq 2$, hence $\dim(V_1 + V_2) = 1$, $V_1 = V_2$ (parallel lines).

If $L_1 \cap L_2$ is not empty, then

$$\dim(L_1 \cap L_2) + \dim(L_1 \circ L_2) = 2.$$

If $L_1 \neq L_2$, then $\dim(L_1 \circ L_2) > 1$, hence $\dim(L_1 \cap L_2) = 0$, $L_1 \cap L_2$ a "point".

6. *Suggestions for an exercise*

The reader will come to a clearer understanding of the content of §§2–4 if he works out by himself the analogue for ordinary additively written abelian groups:

Let R be the additive group of real numbers and Z the subgroup of the ordinary integers. We must first familiarize ourselves with the factor group R/Z. The elements of R/Z are the cosets $a + Z$ with $a \in R$. Two cosets, $a + Z$ and $b + Z$, are equal if and only if $a - b$ is an integer. To simplify notations it shall be understood that we describe $a + Z$ by merely giving a (where a is only defined up to an integer). Such a coset may be multiplied by an integer (this can be done in any additive group) but a product of two cosets is not well defined. One of the main properties of R/Z is that it is "divisible": To an element $a \in R/Z$ and an integer $n > 0$ one can find an element b such that $nb = a$. However, this b is not uniquely defined. Aside from the coset given by a/n one has also the cosets $a/n + i/n$ for $0 \leq i \leq n - 1$ as a possible b.

The reader should now go over our whole exposition on vector spaces, replacing everywhere the word vector space by abelian group, subspace by subgroup and dimension by order of the group. Whenever a plus sign occurs between dimensions it has to be replaced by a product sign. Any reference to the field k has to be dropped. We keep the notations for easier comparison: V now means additive group, $\dim V$ its order. We have no difficulty with the symbol $\langle S \rangle$ but disregard the notion of independence of a set. By a basis of a *finite* abelian group V we mean elements $A_1, A_2, \cdots, A_r$ of V whose orders are $e_1, e_2, \cdots, e_r \in Z$ such that the A_i generate V and such that $m_1A_1 + m_2A_2 + \cdots + m_rA_r = 0$ ($m_i \in Z$) if and only if each m_i is a multiple of e_i. The reader may consult any book on group theory for the proof that a finite abelian group has a basis. The notion of a basis does not work any longer for infinite groups and has to be abandoned. Consequently we can not prove Theorem 1.3 and prove directly that for $U_1 \subset U_2 \subset U_3$ the analogue of equation (1.1), namely

$$\dim U_3/U_1 = \dim U_2/U_1 \cdot \dim U_3/U_2 ,$$

holds. We can now prove Theorem 1.4. In §3 we need only the fact that Hom(V, V') can be made into an abelian group; then we immediately go over to §4.

It is in the analogue of Definition 1.5 that the special group R/Z comes into play. We define $\hat{V} = \text{Hom}(V, R/Z)$ and talk in Definition 1.6 of a pairing of abelian groups W and V into R/Z.

If V is finite with the basis A_i (e_i the order of A_i), let $\varphi \in \hat{V}$. Then $\varphi A_i = a_i \in R/Z$ determines φ. In R/Z the a_i can, however, not be freely selected since $e_i A_i = 0$, and, consequently, $e_i a_i = 0$ (zero element of R/Z). This restricts a_i to an element of the form m/e_i with $0 \leq m \leq e_i - 1$. Within this restriction we have a free choice. This allows us to define a dual basis φ_i by letting $\varphi_i A_j = 1/e_j \cdot \delta_{ij}$. It turns out that the φ_i are a basis of $\hat{V}$ with exactly the same orders e_i. Therefore $\hat{V} \simeq V$ although the isomorphism is not canonical, since it depends on the choice of a basis for V. For a finite V one has, therefore, dim V = dim V.

To handle infinite groups the reader should try to prove the following lemma: Let U be a proper subgroup of V, $X \in V$, $X \notin U$. Let $\varphi \in \hat{U}$. Then φ can be extended to a functional on the group generated by U and the element X consisting of all elements $Y + mX$, $Y \in U$, $m \in Z$. He should try the definition $\varphi(Y + mX) = \varphi(Y) + ma$ with a suitable $a \in R/Z$. How should a be selected so as to make the map well defined? In how many ways can one do it? Can one do it with an $a \neq 0$? Now one uses a transfinite argument to prove that φ can be extended to the whole group V. The reader will have clear sailing up to the end (i.e., Theorem 1.12).

The notion of the dual of a space pervades all of modern mathematics. The reader will meet it again in the theory of Banach spaces and other topics of analysis.

7. *Notions of group theory*

Let G be a group, not necessarily commutative, and S any subset of G.

DEFINITION 1.8. The set of all $x \in G$ for which the set xS is the same as the set Sx is called the *normalizer* N_S of the set S. The set of all $x \in G$ for which $xs = sx$ for all $s \in S$ is called the *centralizer* Z_S of S. The set of all $x \in G$ which commute with all elements of G, in other words the centralizer Z_G of G is called the center of G.

N_S and Z_S are subgroups of G and the center is a commutative invariant (normal) subgroup of G.

Let f be a homomorphism of G into some other group. When is the image group $f(G)$ commutative? $f(a)f(b) = f(b)f(a)$ is equivalent with $f(aba^{-1}b^{-1}) = 1$. If K is the kernel of f it means $aba^{-1}b^{-1} \in K$. The element $aba^{-1}b^{-1}$ is called the commutator of a and b. Its inverse is $bab^{-1}a^{-1}$, the commutator of b and a. The kernel K contains all commutators. Conversely, let H be any subgroup of G which contains all commutators. If $x \in G$ and $h \in H$, then $xh = xhx^{-1}h^{-1} \cdot h \cdot x \in Hx$, hence $xH \subset Hx$ and similarly $Hx \subset xH$. Such an H is automatically invariant and since the kernel H of the canonical map $G \to G/H$ contains all commutators, the factor group G/H will be abelian. The smallest subgroup G' of G which we can form in this way consists of all products of commutators and $G' \subset K$ is another way of expressing the condition we found.

DEFINITION 1.9. The set of all products of commutators of G is called the commutator subgroup G' of G. The factor group G/G' is abelian and the image of G under a homomorphism f will be an abelian group if and only if G' is contained in the kernel K of f. A subgroup H of G which contains G' is necessarily invariant and G/H commutative.

Let c be any element of G and consider the following map $\varphi_c : G \to G$:

$$\varphi_c(x) = cxc^{-1}.$$

We have $\varphi_c(xy) = cxyc^{-1} = cxc^{-1} \cdot cyc^{-1} = \varphi_c(x)\varphi_c(y)$ so that φ_c is a homomorphism.

$$\varphi_c(\varphi_d(x)) = c(dxd^{-1})c^{-1} = cdx(cd)^{-1} = \varphi_{cd}(x),$$

in other words $\varphi_c\varphi_d = \varphi_{cd}$. The map φ_1 is the identity. If $x \in G$ is given, then $x = \varphi_1(x) = \varphi_c(\varphi_{c^{-1}}(x))$ which shows that φ_c is onto. If $\varphi_c(x) = 1$, then $\varphi_{c^{-1}}(\varphi_c(x)) = 1$ or $x = 1$. The kernel is 1, each φ_c is an isomorphism of G onto G, in other words an automorphism of G. This particular type of automorphism φ_c is called an *inner* automorphism of G. Since $\varphi_c\varphi_d = \varphi_{cd}$ and $\varphi_1 = 1$, we see that the inner automorphisms of G form a group I_G.

Consider the map $G \to I_G$ given by $c \to \varphi_c$. It is onto by definition and is a homomorphism. The element c will be in the kernel if $\varphi_c = 1$,

$cxc^{-1} = x$ for all $x \in G$. The kernel of this map is the center of G. We see that

$$I_G \simeq G/Z_G .$$

Let us call two subsets S and T of G equivalent if T is the image of S under some $\varphi_c : T = \varphi_c(S)$. Then $S = \varphi_{c^{-1}}(T)$; and if T and U are equivalent, $U = \varphi_d(T)$, then $U = \varphi_d\varphi_c(S) = \varphi_{dc}(S)$ so that U and S are equivalent. How many sets are equivalent to a given S? We have to decide: when is $\varphi_c(S) = \varphi_d(S)$, or $S = \varphi_{c^{-1}d}(S)$. But $S = \varphi_a(S)$ means $S = aSa^{-1}$ or $Sa = aS$ or $a \in N_S$. Thus we have: $\varphi_c(S) = \varphi_d(S)$ is equivalent to $c^{-1}d \in N_S$, $d \in cN_S$. All elements d of the left coset cN_S will give the same image $\varphi_d(S)$. The number of distinct images $\varphi_a(S)$ is, therefore, equal to the number of left cosets cN_S of N_S. The number $(G : H)$ of left cosets cH which a subgroup H has in G is called the index of H in G. Thus we have proved: The number of distinct sets $\varphi_c(S) = cSc^{-1}$, which we will get when c ranges over G, is equal to the index $(G : N_S)$ of the normaliser of S.

Of special importance is the case where S consists of one element only: $S = \{a\}$. Then each $\varphi_c(S)$ contains only the element cac^{-1}. The elements of G are, therefore, *partitioned* into equivalence classes: a is equivalent to all cac^{-1}. The number of elements in the equivalence class of a is $(G : N_a)$. Which of these equivalence classes contain *one* element only? For such an a we must have $a = cac^{-1}$ for all $c \in G$. They are, therefore, the elements a of the center Z_G of G. Denoting by $\#(S)$ the number of elements in a set S we have, therefore, $\#(Z_G)$ equivalence classes which contain one element only. Counting the number of elements in G by equivalence classes leads to a formula,

$$\#(G) = \#(Z_G) + \sum_a (G : N_a), \tag{1.15}$$

where $\sum_a$ means only rather vaguely "sum over certain a" but where each $(G : N_a) > 1$ since those cases where $(G : N_a) = 1$ have already been counted in $\#(Z_G)$.

Although the following application is not needed, we can not resist the temptation of mentioning it.

Let G be a group whose order $\#(G) = p^r$ $(r \geq 1)$ is a power of a prime p. Then each term of $\sum_a$ is a power of p since $(G : N_a)$ divides p^r. Since each term is > 1, the whole $\sum_a$ is divisible by p. Therefore

$\#(Z_G) = \#G - \sum_a$ is divisible by p. We have the famous theorem that the order of Z_G is > 1, Z_G is not just identity.

If we have a multiplicative group G we may adjoin to it a zero element 0 such that $a \cdot 0 = 0 \cdot a = 0$ for any a in G and also for $a = 0$, and obtain a new set which is not a group but shall be called rather inaccurately a group with 0-element.

DEFINITION 1.10. By a "group with 0 element" we shall mean a union of an ordinary group and an element 0 such that $a \cdot 0 = 0 \cdot a = 0$ for all a in the set.

We have now to describe what we mean by an ordered group: It shall be a group G with an additional binary relation $a < b$ satisfying all laws one would like to have true. Denote by S the set of all $a > 1$. We would like to conclude from $a > 1$, $b > 1$ that $ab > 1$. S will have to be closed under multiplication. One also would like to conclude that $a > 1$ is equivalent with $a^{-1} < 1$ and certainly exclude $a = 1$. Furthermore, we would like to have each element either > 1 or $= 1$ or < 1. This shows that we should postulate that G be the disjoint union $S^{-1} \cup \{1\} \cup S$ where S^{-1} denotes the set of inverses of S. Finally $a > 1$ should imply $ba > b$; multiplying on the right by b^{-1} we would like to get $bab^{-1} > 1$. Let us try this as a definition.

DEFINITION 1.11. A group G is said to be ordered if a set S is singled out with the properties:

1) G is the disjoint union $S^{-1} \cup \{1\} \cup S$,
2) $S \cdot S \subset S$ (closed under multiplication),
3) for any $b \in G$ we have $bSb^{-1} \subset S$.

Applying 3) for b^{-1} we get $b^{-1}Sb \subset S$ or $S \subset bSb^{-1}$ and, therefore, $bSb^{-1} = S$.

We define $a > b$ as meaning $b^{-1}a \in S$. This gives $b \cdot (b^{-1}a)b^{-1} \in S$ or $ab^{-1} \in S$. It is, therefore, irrelevant whether one says $b^{-1}a \in S$ or $ab^{-1} \in S$. Notice that $a > 1$ really means $a \in S$ as we would like it to be.

Given any pair a, b. If $b^{-1}a \in S$, then $a > b$; if $b^{-1}a = 1$, then $a = b$; if $b^{-1}a \in S^{-1}$, then $a^{-1}b \in S$, hence $b > a$ and these possibilities exclude each other. Any two elements are "comparable".

Suppose $a > b$ and $b > c$, then $ab^{-1} \in S$ and $bc^{-1} \in S$, hence $ab^{-1} \cdot bc^{-1} = ac^{-1} \in S$ and we get $a > c$, the transitivity of $>$.

Suppose $a > b$ and let c be any element in G; then $b^{-1}a \in S$ and

$ab^{-1} \varepsilon S$ and consequently $(cb)^{-1}(ca) = b^{-1}a \varepsilon S$ and $ac(bc)^{-1} = ab^{-1} \varepsilon S$ which means that $a > b$ implies $ca > cb$ as well as $ac > bc$. One can multiply an inequality by c.

If $a > b$ and $c > d$ then $ac > ad$ and $ad > bd$, hence by transitivity $ac > bd$. One can multiply two inequalities.

If $a > b$ then $b^{-1} > a^{-1}$, since this means $ab^{-1} \varepsilon S$.

All intuitive laws for inequalities are satisfied.

DEFINITION 1.12. A "group with 0" $= 0 \cup G$ is called ordered if the ordinary group G is ordered in the sense of Definition 1.11 and if 0 is defined to be $<$ than any element of G.

Again all intuitive laws are true.

8. *Notions of field theory*

We have remarked repeatedly that a field k need not have commutative multiplication but the addition is of course commutative. The set of non-zero elements of k forms a multiplicative group which shall be denoted by k^*.

As in any additive group we can multiply an element $a \varepsilon k$ by an ordinary integer $n \varepsilon Z$ (Z shall denote the set of ordinary integers) and get an element $na \varepsilon k$. In any additive commutative group one has the rules $(n + m)a = na + ma$, $n(ma) = (nm)a$, $n(a + b) = na + nb$. In k one sees easily the additional rule

$$(na)(mb) = (nm)(ab).$$

For instance if n and m are > 0 the left side is

$$(a + a + \cdots)(b + b + \cdots)$$

and upon expansion of this product one gets the right side.

Let $a \neq 0$. The question whether $na = 0$ does not depend on a since this is equivalent with $na \cdot a^{-1} = 0$ or $ne = 0$ where e is the unit element of k. This equation does certainly not hold for $n = 1$. If we map the ring Z of integers into k by $n \rightarrow ne$, then this map is a ring homomorphism. If $ne \neq 0$ whenever $n \neq 0$, then it is an isomorphism; k contains in this case an isomorphic replica of Z and, since k is a field, an isomorphic replica of the field Q of rational numbers. A field k of this type is called a field of characteristic 0. Suppose, on the other hand, that the map $n \rightarrow ne$ has a non-zero kernel H, a subgroup of the additive group Z. Such a subgroup $\neq 0$ consists of all multiples

$p\nu$ of the smallest *positive* integer p in H. As remarked earlier $p \neq 1$. If p were not a prime number, then $p = ab$ with positive a, b which are $< p$. Since $p \cdot e = 0$, we would get $(ae)(be) = 0$ which is not true since k is a field. This number p is called the characteristic of a field of this type and is a prime number.

The only distinct images of Z in k are the p elements νe with $0 \leq \nu \leq p - 1$. The $p - 1$ non-zero elements among them are closed under multiplication and, since in a field the cancellation law holds, they form a group. The p elements νe form, therefore, a subfield Q_p of k (which is isomorphic to the field of residue classes Z/pZ of integers modulo p). From now on we will denote the unit element of k by 1, the elements $\nu \cdot 1$ simply by ν with the understanding that ν is to be read modulo the characteristic of k.

If k is a subfield of a field F, we may regard F as a left vector space over k, taking as definition of the vector space operations the ones we already have in F. This space F has a dimension (over k) which is called the left degree of F over k and denoted by $[F : k]$. One could of course also define a right degree; it is an unsolved problem to decide whether any connection exists between these two degrees. We shall stick consistently to the left degree. By a left k-basis $\{a_i\}$ of F one means, therefore, a set of elements in F such that any element of F is a unique finite left linear combination of these basis elements with coefficients in k. We may write $\beta = \sum_i x_i a_i$ where $x_i \in k$ and where $x_i \neq 0$ can hold only for a finite number of indices i.

If F is a subfield of E and k a subfield of F, let $\{\Gamma_j\}$ be a left F-basis of E and $\{\alpha_i\}$ a left k-basis of F. Given an element $A \in E$ we can write $A = \sum_j \beta_j \Gamma_j$ with $\beta_j \in F$ and $\beta_j = \sum_i x_{ij} \alpha_i$ with $x_{ij} \in k$ where only a finite number of the x_{ij} are $\neq 0$. We obtain

$$A = \sum_{i,j} x_{ij} \alpha_i \Gamma_j \, .$$

If, conversely, we have a sum $\sum_{i,j} x_{ij} \alpha_i \Gamma_j$ and call it A, then $A = \sum_j (\sum_i x_{ij} \alpha_i) \Gamma_j$; this A determines the coefficients $\sum_i x_{ij} \alpha_i$ of the Γ_j uniquely and each $\sum_i x_{ij} \alpha_i$ determines the x_{ij} uniquely. This shows that the set $\{\alpha_i \Gamma_j\}$ is a k-basis of E and we have proved the formula

$$[E : k] = [E : F][F : k].$$

If $a \in k$, then we call the set of all $x \in k$ for which $xa = ax$ the normalizer N_a of a. If $x \in N_a$ and $y \in N_a$, then $x \pm y \in N_a$ and $xy \in N_a$.

Should $x \neq 0$, then $xa = ax$ implies $ax^{-1} = x^{-1}a$, hence $x^{-1} \varepsilon N_a$. This proves that N_a is a subfield of k. N_a^* is the group theoretical normalizer of a in the group k^*. The set of all $x \varepsilon k$ such that $xy = yx$ for *all* $y \varepsilon k$ forms also a field Z_k and again we have that Z_k^* is the center of the group k^*. We remark that trivially $Z_k \subset N_a$.

A certain geometric problem is connected with fields k which contain only a finite number s of elements. The characteristic of such a field must be a prime $p > 0$. If k is a subfield of F and $[F : k] = r$, then each element of F is obtained uniquely in the form

$$\sum_{i=1}^{r} x_i \alpha_i , \qquad x_i \epsilon k,$$

where the α_i form a k-basis of F. Hence we see that F contains s^r elements.

Denote by Z_k the center of k and let q be the number of elements in Z_k. Call $n = [k : Z_k]$ and let a be any element of k. Then $Z_k \subset N_a \subset k$ and we introduce the degrees: $d_a = [N_a : Z_k]$ and $e_a = [k : N_a]$. The relation $n = d_a e_a$ shows that d_a is a divisor of n.

We had called q the number of elements in Z_k; q^n and q^{d_a} are, therefore, the number of elements in k, N_a, respectively, and the number of elements in Z_k^*, N_a^*, k^* are $q - 1$, $q^{d_a} - 1$, $q^n - 1$, respectively. We apply now formula (1.15) of §7 to the group $G = k^*$ and obtain a formula which looks like this:

$$q^n - 1 = (q - 1) + \sum_d \frac{q^n - 1}{q^d - 1} \tag{1.16}$$

where $\sum_d$ means vaguely a sum of terms, each one of the form $(q^n - 1)/(q^d - 1)$ and the same term possibly repeated. Indeed, $(G : N_a^*) = (q^n - 1)/(q^{d_a} - 1)$. The d should always be a divisor of n and be $< n$ since each $(G : N_a^*)$ should be > 1.

Our aim is to show that a formula like (1.16) can not exist if $n > 1$. If we succeed, we will have proved $n = 1$ and, therefore, $k = Z_k$. This would mean that k itself is commutative. To give the proof we first need some facts about the so-called cyclotomic polynomials.

It is well known that the polynomial $x^n - 1$ can be factored in the field of complex numbers:

$$x^n - 1 = \prod_{\epsilon} (x - \epsilon) \tag{1.17}$$

where ϵ ranges over the n-th roots of unity: $\epsilon^n = 1$. If d is the *precise* order of ϵ, then ϵ is called a *primative* d-th root of unity. Every d-th root of unity will appear among the n-th roots of unity if d is a divisor of n. Let us define

$$\Phi_d(x) = \prod (x - \epsilon)$$

where ϵ shall range only over the *primitive* d-th roots of unity. Grouping the factors of (1.17) we find

$$x^n - 1 = \prod_{d|n} \Phi_d(x) \tag{1.18}$$

where d ranges over all divisors of n. These polynomials $\Phi_n(x)$ are the cyclotomic polynomials. The $\Phi_n(x)$ have obviously highest coefficient 1. We contend now that all coefficients of $\Phi_n(x)$ are integers. Since $\Phi_1(x) = x - 1$, this contention is true for $n = 1$ and we may assume it is proved for all $\Phi_d(x)$ with $d < n$. We know then that (1.18) has the form

$$x^n - 1 = \Phi_n(x) \cdot f(x)$$

where $f(x)$ is the product of the factors $\Phi_d(x)$ with $d < n$ and $d|n$. Therefore $f(x)$ has integral coefficients and its highest coefficient is 1. The desired polynomial $\Phi_n(x)$ can, therefore, be obtained by dividing $x^n - 1$ by $f(x)$. Remembering the way this quotient is computed and the fact that the highest coefficient of $f(x)$ is 1, we see that $\Phi_n(x)$ has integral coefficients.

Let now d be a divisor of n but $d < n$. Then

$$x^d - 1 = \prod_{\delta|d} \Phi_\delta(x);$$

each term $\Phi_\delta(x)$ will appear as a factor on the right side of (1.18) since $\delta|n$. But $\delta \neq n$; in the quotient $(x^n - 1)/(x^d - 1)$ the polynomial $\Phi_n(x)$ will still be one of the factors. Thus we see that $\Phi_n(x)$ divides $x^n - 1$ as well as $(x^n - 1)/(x^d - 1)$,

$$x^n - 1 = \Phi_n(x)f(x), \qquad \frac{x^n - 1}{x^d - 1} = \Phi_n(x)g(x),$$

and both $f(x)$ and $g(x)$ will have integral coefficients. If we set $x = q$, we see that the integer $\Phi_n(q)$ divides the two integers $q^n - 1$ and $(q^n - 1)/(q^d - 1)$. With this information we turn to (1.16) and can conclude that the integer $\Phi_n(q)$ must divide $q - 1$.

Now we estimate the size of the integer $\Phi_n(q)$. It is a product of terms $q - \epsilon$. The absolute value of $q - \epsilon$ is the distance of the point ϵ on the unit circle and the point $q \geq 2$ on the real axis. Each factor is certainly ≥ 1, even $\geq q - 1$, in absolute value and can be equal to $q - 1$ only for $\epsilon = 1$. This case $\epsilon = 1$ does not occur if $n > 1$ since ϵ is a primitive n-th root of unity. Thus certainly $|\Phi_n(q)| > q - 1$ if $n > 1$. We see clearly that $\Phi_n(q)$ could not divide $q - 1$. Thus we have proved the celebrated theorem of Wedderburn:

THEOREM 1.14. *Every field with a finite number of elements is commutative.*

DEFINITION 1.13. Let f be a map of a field k into some field F which is one-to-one into and is a homomorphism for addition. If f satisfies

$$f(ab) = f(a)f(b) \tag{1.19}$$

for all $a, b \in k$ we call f an isomorphism of k into F. If f satisfies

$$f(ab) = f(b)f(a) \tag{1.20}$$

for all $a, b \in k$ we call f an antiisomorphism of k into F.

Let us remark that it suffices to assume that f is a homomorphism for addition, does not map all of k onto 0, and satisfies either (1.19) or (1.20). Indeed, if $f(a) = 0$ for a single $a \neq 0$, then it would already follow that $f(ak) = 0$. Since k is a field, $ak = k$, which contradicts our assumption.

Hua has discovered a beautiful theorem which has a nice geometric application:

THEOREM 1.15. *If σ is a map of a field k into some field F which satisfies the following conditions:*

1) *σ is a homomorphism for addition,*

2) *for $a \neq 0$ we have $\sigma(a^{-1}) = (\sigma(a))^{-1}$; i.e., we assume that σ maps the inverse of an element onto the inverse of the image,*

3) *$\sigma(1) = 1$;*

then σ is either an isomorphism or an antiisomorphism of k into F.

REMARK. Suppose σ satisfies only conditions 1) and 2). Set $a = 1$ in 2), then $x = \sigma(1)$ satisfies $x = x^{-1}$, that is $x^2 - 1 = (x - 1)(x + 1) = 0$. Thus, if σ does not satisfy condition 3), $\sigma(1) = -1$. If we put $\tau(a) = -\sigma(a)$, then τ will satisfy all three conditions which means that σ is either the negative of an isomorphism or the negative of an antiisomorphism.

Proof of the theorem: Instead of $\sigma(a)$ we shall write a^σ. Condition 2) reads now $(a^{-1})^\sigma = (a^\sigma)^{-1}$ for which we can write unambiguously $a^{-\sigma}$. Since $a \neq 0$ implies $a^\sigma \cdot a^{-\sigma} = 1$, we have $a^\sigma \neq 0$ which shows that the additive kernel of σ is 0 and our map, therefore, one-to-one into.

We first establish an identity:

Assume $a, b \in k$; $a, b \neq 0$ and $a^{-1} \neq b$. Then the expression $a^{-1} + (b^{-1} - a)^{-1}$ is well defined. Let us factor out a^{-1} to the left and $(b^{-1} - a)^{-1}$ to the right:

$$a^{-1} + (b^{-1} - a)^{-1} = a^{-1}((b^{-1} - a) + a)(b^{-1} - a)^{-1}$$
$$= a^{-1}b^{-1}(b^{-1} - a)^{-1}.$$

We can, therefore, take the inverse:

$$(a^{-1} + (b^{-1} - a)^{-1})^{-1} = (b^{-1} - a)ba = a - aba.$$

Thus we have

$$a - (a^{-1} + (b^{-1} - a)^{-1})^{-1} = aba.$$

If we apply σ to the left side, conditions 1) and 2) allow us to interchange σ each time with the operation of taking the inverse and we will end up with an expression like the one we started with, but a replaced by a^σ and b replaced by b^σ. The left side becomes $a^\sigma b^\sigma a^\sigma$ and we have shown that

$$(aba)^\sigma = a^\sigma b^\sigma a^\sigma. \tag{1.21}$$

(1.21) is also true if a or b are 0. If $a^{-1} = b$, then $ab = 1$, the left side of (1.21) is a^σ. But $a^\sigma b^\sigma a^\sigma = a^\sigma a^{-\sigma} a^\sigma = a^\sigma$ which shows that (1.21) is true for all $a, b \in k$.

Set $b = 1$ in (1.21);

$$(a^2)^\sigma = (a^\sigma)^2. \tag{1.22}$$

Replace a in (1.22) by $a + b$:

$$(a^2 + ab + ba + b^2)^\sigma = (a^\sigma)^2 + a^\sigma b^\sigma + b^\sigma a^\sigma + (b^\sigma)^2.$$

If we use (1.22) and condition 1) we finally obtain

$$(ab)^\sigma + (ba)^\sigma = a^\sigma b^\sigma + b^\sigma a^\sigma. \tag{1.23}$$

Now comes the main trick. Let $a, b \neq 0$ and consider

$$((ab)^\sigma - a^\sigma b^\sigma)(ab)^{-\sigma}((ab)^\sigma - b^\sigma a^\sigma). \tag{1.24}$$

Multiplying out (1.24) becomes equal to

$$(ab)^\sigma - b^\sigma a^\sigma - a^\sigma b^\sigma + a^\sigma b^\sigma (ab)^{-\sigma} b^\sigma a^\sigma. \tag{1.25}$$

Use (1.21) on $b^\sigma (ab)^{-\sigma} b^\sigma = (b(ab)^{-1}b)^\sigma$, write $a^\sigma (b^\sigma (ab)^{-\sigma} b^\sigma) a^\sigma = a^\sigma (b(ab)^{-1}b)^\sigma a^\sigma$ and use (1.21) again. We obtain

$$(a \cdot b(ab)^{-1} b \cdot a)^\sigma = (ba)^\sigma.$$

Thus (1.24) is equal to

$$(ab)^\sigma - b^\sigma a^\sigma - a^\sigma b^\sigma + (ba)^\sigma$$

which is zero by (1.23). Thus the product (1.24) is zero and one of the factors must vanish.

We now know that

$$(ab)^\sigma = \begin{cases} a^\sigma b^\sigma \\ \text{or} \\ b^\sigma a^\sigma \end{cases} \tag{1.26}$$

which is much nearer to what we want to show. For a or $b = 0$, (1.26) is trivially true.

We ask whether the following situation could occur in k: can four elements a, b, c, d exist in k such that

$$\begin{aligned} (ab)^\sigma &= a^\sigma b^\sigma \neq b^\sigma a^\sigma, \\ (cd)^\sigma &= d^\sigma c^\sigma \neq c^\sigma d^\sigma? \end{aligned} \tag{1.27}$$

We shall derive a contradiction from this assumption.

Let x be any element in k and use (1.26) on a and $b + x$:

$$(a(b+x))^\sigma = \begin{cases} a^\sigma (b+x)^\sigma = a^\sigma b^\sigma + a^\sigma x^\sigma \\ \text{or} \\ (b+x)^\sigma a^\sigma = b^\sigma a^\sigma + x^\sigma a^\sigma. \end{cases} \tag{1.28}$$

The left side is $(ab + ax)^\sigma = a^\sigma b^\sigma + (ax)^\sigma$. If the first case of (1.28) happens we get $(ax)^\sigma = a^\sigma x^\sigma$. If the second case of (1.28) happens remember that $a^\sigma b^\sigma \neq b^\sigma a^\sigma$ so that certainly $(ax)^\sigma \neq x^\sigma a^\sigma$. This means, by (1.26), that $(ax)^\sigma = a^\sigma x^\sigma$. We have, therefore, always

$(ax)^\sigma = a^\sigma x^\sigma$. The same method is used on the expressions $((a + x)b)^\sigma$, $(c(d + x))^\sigma$ and $((c + x)d)^\sigma$. Collecting all four cases the results are:

$$(1.29) \qquad (ax)^\sigma = a^\sigma x^\sigma,$$

$$(1.30) \qquad (xb)^\sigma = x^\sigma b^\sigma,$$

$$(1.31) \qquad (cx)^\sigma = x^\sigma c^\sigma,$$

$$(1.32) \qquad (xd)^\sigma = d^\sigma x^\sigma.$$

Set $x = d$ in (1.29) and $x = a$ in (1.32); set also $x = c$ in (1.30) and $x = b$ in (1.31). We obtain

$$(1.33) \qquad a^\sigma d^\sigma = d^\sigma a^\sigma \quad \text{and} \quad c^\sigma b^\sigma = b^\sigma c^\sigma.$$

Finally:

$$(1.34) \qquad ((a + c)(b + d))^\sigma = \begin{cases} (a + c)^\sigma(b + d)^\sigma = a^\sigma b^\sigma + a^\sigma d^\sigma + c^\sigma b^\sigma + c^\sigma d^\sigma \\ \text{or} \\ (b + d)^\sigma(a + c)^\sigma = b^\sigma a^\sigma + d^\sigma a^\sigma + b^\sigma c^\sigma + d^\sigma c^\sigma. \end{cases}$$

A direct computation of the left side gives:

$$(ab)^\sigma + (ad)^\sigma + (cb)^\sigma + (cd)^\sigma = a^\sigma b^\sigma + a^\sigma d^\sigma + c^\sigma b^\sigma + d^\sigma c^\sigma.$$

The first possibility on the right of (1.34) would give $c^\sigma d^\sigma = d^\sigma c^\sigma$ which contradicts (1.27). Using (1.33) for the second possibility we get $a^\sigma b^\sigma = b^\sigma a^\sigma$ and this contradicts (1.27) again. Our theorem is finally established; the fact that (1.27) can not happen obviously means that in (1.26) we either have the first possibility for all $a, b \in k$ or else the second one for all $a, b \in k$.

9. *Ordered fields*

DEFINITION 1.14. A field k is said to be ordered, if, first of all, it is ordered as an additive group. Rewriting Definition 1.11 in the additive notation (the third condition in 1.11 is not necessary since addition is commutative) this means that a set P of so-called "positive" elements is singled out such that

1) $k = -P \cup \{0\} \cup P$ (disjoint),
2) $P + P \subset P$ (P is closed under addition).

To these two additive conditions we add a multiplicative one:

3) $P \cdot P \subset P$ (a product of positive elements is positive).

We define now $a > b$ by $a - b \in P$, mention again that $a > 0$ is equivalent with $a \in P$ and, therefore, can use all consequences of Definition 1.11—of course in the additive form—among them for example that $a > b$ implies $-b > -a$. Let us now derive multiplicative consequences.

a) If $c \in P$ and $a > b$, then $a - b \in P$. Therefore, $c(a - b) \in P$ and $(a - b)c \in P$, hence $ca > cb$ and $ac > bc$. If $-c \in P$, then $-c(a - b) \in P$, hence $cb > ca$ and $bc > ac$ (inequalities are reversed if $0 > c$).

b) If $a > 0$, then $a \cdot a > 0$. If $0 > a$ multiplication by a reverses the inequality, hence also in this case $a \cdot a > 0$. Squares of non-zero elements are positive and especially $1 = 1^2 > 0$. Therefore $0 > -1$. We can write $c^{-1} = c \cdot (c^{-1})^2$ and see that $c > 0$ implies $c^{-1} > 0$, $0 > c$ implies $0 > c^{-1}$.

Multiplying on the left by $c \neq 0$ and on the right by c^{-1} keeps the inequality: $a > b$ implies $cac^{-1} > cbc^{-1}$. The special cases $b = 0$ and $b = 1$ show that $cPc^{-1} \subset P$ and, if S is the set of elements $a > 1$, that $cSc^{-1} \subset S$. The set P is an invariant multiplicative subgroup of k^*; $a > b$ is equivalent with $ab^{-1} > 1$, hence with $ab^{-1} \in S$. If $a \in P$ and neither > 1 not $= 1$, then $1 > a$. Multiplying by a^{-1} gives $a^{-1} > 1$ or $a^{-1} \in S$.

The group P is the disjoint union $S^{-1} \cup \{1\} \cup S$ and our ordering of k induces an ordering of the multiplicative group P.

c) Since $1 \in P$, any sum of terms 1 is in P which shows that the characteristic of k must be 0. The ordinary positive integers are also positive in k. The ordering of k induces, therefore, on the subfield Q of rational numbers the usual ordering. This proves also that the field of rational numbers can only be ordered in the usual way.

In some geometric problems the ordering of a field arises in a quite different manner. We are led to another definition.

DEFINITION 1.15. A field k is said to be weakly ordered if a binary relation $a < b$ is defined in k such that the usual ordering properties hold. The connection to field operations presupposes, however, much weaker axioms:

1) For any fixed $a \in k$ the ordering of k shall either be preserved or reversed under the map $x \to x + a$. (Notice that some a may preserve it, others may reverse it.)

2) For any fixed $a \in k^*$ the ordering of k shall either be preserved

or reversed under the map $x \to xa$. (Nothing is said about a map $x \to ax$.)

If we say elements are in the arrangement $a_1, a_2, \cdots, a_n$ we mean that either $a_1 < a_2 < \cdots < a_n$ or $a_1 > a_2 > \cdots > a_n$. The axioms say that the maps preserve the arrangements.

There is one freak case: the field with two elements 0, 1 has obviously only one arrangement so that any one-to-one map preserves it; the field can be weakly ordered. Suppose from now on that k has at least three elements. We prove certain facts about such a weakly ordered field.

1) k can not have the characteristic 2. Let indeed $0, a, b$ be three distinct elements of k and suppose that k is weakly ordered and has characteristic 2.

a) Our elements are arranged as $0, a, b$. If we add a we get the arrangement $a, 0, a + b$ and if we add b we get $b, a + b, 0$. These two arrangements combined give $a, 0, a + b, b$ and this contradicts the original one.

b) The arrangement is $a, 0, b$. Adding a and b we get $0, a, a + b$ and $a + b, b, 0$ whose combination gives either $0, a, b, a + b$ or $0, b, a, a + b$ contradicting the assumption.

These two types exhaust the possibilities.

2) k has now a characteristic $\neq 2$. Let $a \in k^*$ and consider the possible arrangements of $0, a/2$ and $-a/2$.

a) The arrangement is $-a/2, 0, a/2$. Adding $-a/2$ and $a/2$ we get the arrangements $-a, -a/2, 0$ and $0, a/2, a$ which, together with the original one give

$$-a, -\frac{a}{2}, 0, \frac{a}{2}, a.$$

b) The arrangement is $0, -a/2, a/2$. Adding $-a/2$ and $a/2$ gives $-a/2, -a, 0$ and $a/2, 0, a$. Combined with the original one we get $a, 0, -a, -a/2, a/2$. (The case $0, a/2, -a/2$ is just a sign change in a.)

Both these hypotheses lead to the arrangement

$$-a, 0, a$$

which must, therefore, hold in any field k for any non-zero element. Adding a we also get $0, a, 2a$.

3) Suppose two elements are in the arrangement

$$0, a, b.$$

Adding a we get a, $2a$, $a + b$. Combining it with 0, a, $2a$ we see that we also have 0, a, $a + b$. Thus we have the following rule: If two elements a and b are on the same side of 0, then their sum is also on this same side.

Denote now by P the set of all elements which are on the same side of 0 as the element 1. Then we know that k is the disjoint union

$$-P \cup \{0\} \cup P$$

and that $P + P \subset P$. These are the first two axioms of an ordered field. It is worth while noticing that we used only condition 1) of Definition 1.15.

Suppose $a, b \in P$. Then a is on the same side of 0 as 1. Multiplying by b we get that ab is on the same side of 0 as b. But this means $ab \in P$.

The field k is ordered. We see that the maps $x \to ax$ (for $a \neq 0$) also either preserve or reverse the ordering.

THEOREM 1.16. *If a field k is weakly ordered and has more than two elements, then it is an ordered field.*

Hilbert has constructed an ordered non-commutative field and we shall present his example.

Let F be any field and σ an automorphism of F. We denote by a^σ the image of an element $a \in F$ under σ and by a^{σ^i} its image under the automorphism σ^i.

We construct an extension field k of F consisting of all formal power series

$$\sum_{i=-\infty}^{+\infty} a_i t^i$$

in a variable t which contain only a finite number of non-zero coefficients with a negative i; in other words we assume that for each power series an integer N can be found such that $a_i = 0$ for $i < N$.

The addition in k is defined as usual but the multiplication is not the ordinary one. To multiply two power series one first multiplies termwise:

$$\sum_i a_i t^i \cdot \sum_j b_j t^j = \sum_{i,j} a_i t^i b_j t^j.$$

But now one does not assume that t^i can be interchanged with b_j; instead of an ordinary interchange one uses the rule

$$t^i b = b^{\sigma^i} t^i$$

which for our product produces the series

$$\sum_{i,j} a_i b_j^{\sigma^i} t^{i+j}.$$

Finally one collects terms with the same power of t. The coefficient c_r of t^r will, therefore be

$$c_r = \sum_{i+j=r} a_i b_j^{\sigma^i}.$$

It is a good exercise to make all this rigorous. We indicate the steps and leave the details to the reader. Since a power series is given by the coefficients we may just as well say that the elements of k are functions $f(i)$ on the integers with value in F, thinking of the value $f(i)$ as the coefficient of t^i. We consider only functions f whose value $f(i) = 0$ whenever $i < N$ (some integer depending on f). The addition of functions is defined as usual but the product fg shall be the function whose value on i is given by

$$(fg)(i) = \sum_{\nu+\mu=i} f(\nu)(g(\mu))^{\sigma^\nu}.$$

One has to verify that the sum on the right side has only a finite number of non-zero terms and also that $(fg)(i) = 0$ whenever $i < N + M$; N is the integer belonging to f and M the one belonging to g.

It is easy to show that the two distributive laws and also the associative law hold. Since we wish to consider k as extension of F we have to map F isomorphically into k. The naive description by power series suggests how to do it. An element $a \in F$ is mapped onto the function $\bar{a}$ which satisfies $\bar{a}(0) = a$ and $\bar{a}(i) = 0$ for $i \neq 0$. One verifies that $a \to \bar{a}$ is an isomorphism of F into the ring k. The image $\bar{1}$ of the unit element of F turns out to be the unit of k. If $f \in k$, then $\bar{a}f$ is the function whose value at i is $a \cdot f(i)$, whereas $f\bar{a}$ has at i the value $f(i) \cdot a^{\sigma^i}$.

One wishes also to get the power t^m. To this effect one denotes by t_m the function which has value 1 at m and value 0 at all other integers. Now one proves $t_m t_n = t_{m+n}$ and $t_0 = \bar{1}$, the unit element.

Therefore $t_1^m = t_m$ and abbreviating t_1 by t we see that we can write $t_m = t^m$.

Now one can prove

$$t^m \bar{a} = \overline{a^{\sigma^m}}\, t^m.$$

One begins to drop the bar on top of $\bar{a}$, identifying $\bar{a}$ with a.

If $f \in k$, then ft^m is really a shift by m. One easily verifies the formulas

$$(ft^m)(i) = f(i - m) \quad \text{and} \quad (t^m f)(i) = (f(i - m))^{\sigma^m}.$$

We wish to show that k is a field, i.e., that a non-zero element has an inverse. If $f \neq 0$, then by taking a suitable t^m and a suitable $a \in F$, the new function $g = aft^m$ will satisfy $g(0) = 1$ and $g(i) = 0$ for $i < 0$ (its power series has no negative terms and begins with 1). If one can show that g has an inverse, then f will have the inverse $t^m g^{-1} a$ (since $f = a^{-1} g t^{-m}$). One tries a hypothetical inverse h of the same type, $h(0) = 1$, $h(i) = 0$ for $i < 0$. For $h(1), h(2), \cdots$ one finds equations (starting with $h(0) = 1$)

$$h(1)g(0)^{\sigma} + h(0)g(1) = 0,$$
$$h(2)g(0)^{\sigma^2} + h(1)g(1)^{\sigma} + h(0)g(2) = 0,$$
$$h(3)g(0)^{\sigma^3} + h(2)g(1)^{\sigma^2} + h(1)g(2)^{\sigma} + h(0)g(2) = 0,$$
$$\cdots\cdots\cdots\cdots\cdots\cdots\cdots\cdots\cdots\cdots\cdots\cdots\cdots\cdots$$

which allow successively to compute $h(1), h(2), \cdots$. The inverse will be a left inverse: $f^{-1}f = 1$. But mere group theory shows that then $ff^{-1} = 1$. Our ring k is an extension field of F.

At this point one can safely return to the naive power series description of k. If $f = \sum_i a_i t^i \in k$ and if $f \neq 0$, then there exists an integer N such that $a_i = 0$ if $i < N$ but $a_N \neq 0$. We shall call this a_N the lowest term of f. If a_N is the lowest term of f and b_M the lowest term of g, then $a_N b_M^{\sigma^N}$ will be the lowest term of fg.

It is now time to talk of the ordering of k. We assume that F is an ordered field and try to define a set P of elements of k which shall induce an ordering of k. Let P consist of those $f \neq 0$ whose lowest term a_N is positive in the ordering of F: $a_N > 0$. The condition $k = -P \cup \{0\} \cup P$ (disjoint) is obviously satisfied; the rule that the sum of two elements in P lies also in P is not difficult to verify. There remains the product rule. Here we obviously have to assume of our automorphism that $a > 0$ implies $a^{\sigma} > 0$. Suppose that σ has this property; then $a < 0$ will also imply that $a^\sigma < 0$. It follows that $a > 0$ *is implied* by $a^\sigma > 0$ which shows that σ^{-1} has our property. We needed σ^{-1} in order to be sure that all powers of σ have the same property. Now we can prove the product rule. If the lowest terms a_N, respectively, b_M of f, respectively, g are positive, then the lowest term $a_N b_M^{\sigma^N}$ of fg will be positive. We see that k is ordered.

Is k non-commutative? We have $ta = a^\sigma t$. As soon as σ is not the identity automorphism our field k will be non-commutative.

This reduces the construction problem a little. We must find a field F (commutative or not) which is ordered and which has an automorphism $\sigma \neq 1$ such that σ preserves positivity. One can do this again by power series. Let Q be the field of rational numbers ordered in the usual fashion and F the field of power series in a variable x:

$$\sum_i a_i x^i,$$

but this time with the ordinary multiplication of power series (hence with identity as automorphism). Let us order F by the lowest term; this is possible since the identity automorphism preserves the positivity in Q trivially. We must find an automorphism $\sigma \neq 1$ of F which preserves positivity. It is the substitution $x \to 2x$, or, more precisely:

$$f = \sum_i a_i x^i \to f^\sigma = \sum_i a_i 2^i x^i.$$

To show that σ is an automorphism of F is a rather trivial exercise. That it preserves the sign of the lowest term is completely obvious. This finishes the construction of k.

THEOREM 1.17. *There do exist ordered non-commutative fields.*

An ordered field has characteristic 0 and contains, therefore, the field Q of rational numbers. Since the field Q can be ordered only in the usual way, the ordering of k is in agreement with the usual ordering of Q.

DEFINITION 1.16. An ordered field k is said to be archimedean if for any $a \in k$ one can find an integer n such that $a < n$.

Let k be archimedean and $a \in k$. There are integers n, m such that $a < n$ and $-a < m$; then $-m < a < n$. Every element of k lies between integers. Consider now the set C of all rational numbers $> a$ and the set D of all rational numbers $\leq a$. Neither C nor D

will be empty and these sets form a Dedekind cut in the rational numbers. Denote by $f(a)$ the real number defined by this cut. We have a map $f : k \to R$ where R is the field of real numbers and will show that f is one-to-one into. Let $a \neq b$, say $b > a$; we shall construct a rational number which lies strictly between a and b and thereby prove $f(a) \neq f(b)$. There is an integer $m > (b - a)^{-1}$. Since $b - a > 0$ we obtain $mb - ma > 1$. Among the integers n which satisfy $n > ma$ there is a smallest one. For this n we have $ma \geq n - 1 > n - (mb - ma) = n - mb + ma$ and get $mb > n$. The inequality

$$mb > n > ma$$

shows $m > 0$ and $b > n/m > a$. We conclude $f(b) > f(a)$. From the properties of Dedekind cuts one deduces that f is an isomorphism of k into R; f preserves the ordering as our previous discussion shows.

THEOREM 1.18. *An archimedean field is commutative, as a matter of fact, isomorphic to a subfield of the real numbers in its natural ordering.*

We saw that Q admits only one ordering. The field R of real numbers admits also only one ordering. Indeed squares of non-zero elements are positive in any ordering. Let $a \in R$ be positive in some ordering of R. Then $-a$ can not be a square which means that $a > 0$ in the ordinary sense. Conversely if $a > 0$ in the ordinary sense then a is a square and, therefore, positive in the given ordering.

Let now σ be an automorphism of R; σ carries squares into squares and preserves, therefore, the positivity in R. If $a < b$, then $a^\sigma < b^\sigma$. The automorphism σ induces on Q the identity. Let a be a given element of R. It is uniquely described by the Dedekind cut which it produces among the rational numbers. Since σ preserves inequalities and leaves all rational numbers fixed, a^σ will have the same Dedekind cut as a. But this means $a^\sigma = a$.

THEOREM 1.19. *The field R of real numbers can be ordered only in the usual way. Every automorphism of R is the identity.*

10. *Valuations*

DEFINITION 1.17. Let k be a field and G an ordered multiplicative group with 0-element. We shall assume both k and G to be commutative. A map of k into G, denoted by $a \to |a|$ (read as "absolute value of a") is called a valuation of k if it satisfies the following conditions:

1) $|a| = 0$ if and only if $a = 0$,
2) $|ab| = |a| \cdot |b|$,
3) $|a + b| \leq \text{Max}(|a|, |b|)$.

REMARK. Condition 3) may be replaced by the rule

3a) $|a| \leq 1$ implies $|1 + a| \leq 1$.

Proof: Conditions 1) and 2) imply $|1| = 1$ so that 3) implies 3a). Suppose 3a) is true. Condition 3) certainly holds if $a = b = 0$. Otherwise assume that $|a| \leq |b|$. Then $|a/b| \leq 1$ and 3a) implies $|1 + a/b| \leq 1$. Multiplying both sides by $|b|$ we get 3).

All we shall really need are some examples of valuations. They are easily constructed by means of another notion.

DEFINITION 1.18. Let k be a commutative field. A subring R is said to be a valuation ring if for every non-zero element of k either a or a^{-1} lies in R.

The elements of k for which both a and a^{-1} are in R form a multiplicative group U, called the group of units. Those non-zero elements for which a^{-1} but not a is in R form a set S, and those for which a, but not a^{-1}, is in R form the set S^{-1} consisting of the inverses of the elements in S. Clearly,

$$k = \{0\} \cup S^{-1} \cup U \cup S,$$

and this union is disjoint. The ring R is:

$$R = \{0\} \cup S^{-1} \cup U.$$

If $a, b \in R$ and $ab \in U$, then $a^{-1}b^{-1} \in R$, hence $a^{-1} = b \cdot (a^{-1}b^{-1}) \in R$ and $b^{-1} = a(a^{-1}b^{-1}) \in R$ which shows $a \in U$ and $b \in U$. This proves $S^{-1} \cdot S^{-1} \subset S^{-1}$ and $S^{-1}U \subset S^{-1}$; consequently $SS \subset S$ and $SU \subset S$.

The cosets of U form the set

$$k/U = \{0\} \cup S^{-1}/U \cup \{U/U\} \cup S/U$$

and this union is disjoint. k/U is a group with 0-element, S/U a semi-group. We may, therefore, consider k/U as an ordered group with 0-element.

A valuation is now easily constructed: the map $a \to |a|$ shall merely be the canonical map of k onto k/U, in other words

$$|a| = aU.$$

Conditions 1) and 2) are trivially satisfied. The inequality $|a| \leq 1$ means aU is either in S^{-1}/U or in $U/U = 1$; this shows that $|a| \leq 1$ is equivalent with $a \varepsilon R$ and implies trivially $1 + a \varepsilon R$ and consequently $|1 + a| \leq 1$.

The case $R = k$ is not excluded. S is then the empty set and k/U has only two elements, 0 and $U/U = 1$. We call this valuation the trivial valuation.

Now our examples:

1) Let $k = Q$—the field of rational numbers—and p a given fixed prime. Let R be the set of those elements of Q which can be written with a denominator relatively prime to p. R is obviously a valuation ring. U consists of those fractions whose numerator and denominator are prime to p. If $a \neq 0$, then $|a| = aU = p^i U$ where p^i is the contribution of p to a if we factor a into prime powers; $|a| \leq 1$ means $a \varepsilon R$, hence $i \geq 0$. If $|a| = p^i U$ and $|b| = p^j U$, then $|a| \leq |b|$ implies $|a/b| = p^{i-j} U \leq 1$ or $i - j \geq 0$, $i \geq j$. An element of Q is "small" if it is divisible by a high power of p.

This valuation of Q is called the p-adic valuation.

2) Let k be an ordered field. An element of k is said to be finite if it lies between two integers; if it does not, we may call it infinite. One sees easily that the set of all finite elements forms a valuation ring R. The corresponding set S consists of the infinite elements of k, the set S^{-1} may be called the set of infinitely small elements; $|a| \leq 1$ means now that a is finite.

This valuation will be trivial if and only if k is archimedean. If k is not archimedean, then $| a |$ gives a classification of the infinitely large or small elements "according to size".

CHAPTER II

Affine and Projective Geometry

1. Introduction and the first three axioms

We are all familiar with analytic geometry where a point in a plane is described by a pair (x, y) of *real* numbers, a straight line by a linear, a conic by a quadratic equation. Analytic geometry enables us to reduce any elementary geometric problem to a mere algebraic one. The intersection of a straight line and a circle suggests, however, enlarging the system by introducing a new plane whose points are pairs of *complex* numbers. An obvious generalisation of this procedure is the following. Let k be a given field; construct a plane whose "points" are the pairs (x, y) of elements of k and define lines by linear equations. It is of course comparatively easy to work out the geometric theorems which would hold in such a geometry.

A much more fascinating problem is, however, the converse. Given a plane geometry whose objects are the elements of two sets, the set of points and the set of lines; assume that certain axioms of geometric nature are true. Is it possible to find a field k such that the points of our geometry can be described by coordinates from k and the lines by linear equations?

We shall first state the needed axioms in a loose fashion. The first two are the incidence axioms: there is exactly one line through two distinct points and exactly one parallel to a given line through a given point. The third axiom eliminates freak cases of geometries which are easily enumerated and merely assures us of the existence of enough points and lines. For a while we shall work only with these three axioms and prepare the ground for the fourth (and last) which is the most interesting one.

One introduces certain symmetries of the geometry called dilatations. They are (except for freaks) one-to-one maps of the plane onto itself which move all points of a line into points of a parallel line.

The identity map is such a dilatation and with only our three axioms at ones disposal, one should not expect the geometry to have any others. This is where the fourth axiom comes into play. It is split into two parts, 4a and 4b. Axiom 4a postulates the existence of a translation (a special kind of dilatation) which moves a given point into any other given point. With this axiom 4a alone one can already construct a certain field k. Axiom 4b shall now ensure the existence of enough dilatations of the remaining kind (which resemble the transformations of similarity in the euclidean plane). Using axiom 4b one can show that the points can be described by coordinates from k and the lines by linear equations.

One can, therefore, conclude that one can coordinatize any geometry which contains enough symmetries.

We shall now start with a precise formulation of the axioms.

We are given two sets: a set of "points" and a set of "lines". We are also given a basic binary relation between a given point P and a given line l: "P lies on l" (which may or may not be true for the given pair P and l). All the axioms can be expressed in terms of this one basic relation. However it is clear that the language would become intolerably clumsy if one would actually do this. We shall, therefore, use obvious synonyms for the binary relation as "P is on l" or "l contains P" or "l goes through P". If P lies on both l and m we may say that l and m meet in P and if P is the only point on both l and m, then we say that "l and m intersect in P" or that "P is the intersection of l and m". At the beginning we shall be a little stricter in our language until the reader has learned to replace the synonyms by the basic relation, but we will soon relax.

For a similar reason we did not draw figures in the first paragraphs of this chapter. The reader will, nevertheless, get a better understanding of the proofs if he makes simple sketches for himself.

Our first task is now to define what is meant by "parallelism".

Definition 2.1. If l and m are two lines such that either $l = m$ or that no point P lies on both l and m, then we call l and m parallel and write $l \parallel m$. If l and m are not parallel, we write $l \nparallel m$.

If $l \nparallel m$, then there is at least one point P which lies on both l and m.

Axiom 1. Given two *distinct* points P and Q, there exists a *unique* line l such that P lies on l and Q lies on l. We write $l = P + Q$.

If $l \nparallel m$, then there is exactly one point P which lies on both l and m.

Indeed, if there were two such points, then, by axiom 1, $l = m$, hence $l \parallel m$.

AXIOM 2. Given a point P and a line l, there exists one and only one line m such that P lies on m and such that $m \parallel l$.

THEOREM 2.1. *"Parallel" is an equivalence relation.*

Proof: It is obviously reflexive and symmetric. To prove the transitivity assume $l_1 \parallel l_2$ and $l_2 \parallel l_3$. If there is no point on both l_1 and l_3, then $l_1 \parallel l_3$. If there is a point P on both l_1 and l_3, then by axiom 2 (since $l_1 \parallel l_2$ and $l_3 \parallel l_2$) we have $l_1 = l_3$, hence again $l_1 \parallel l_3$.

DEFINITION 2.2. An equivalence class of parallel lines is called a pencil of parallel lines.

THEOREM 2.2. *Suppose that there exist three distinct pencils, π_1, π_2, and π_3 of parallel lines. Then any pencil π contains the same number of lines and this number is equal to the number of points on any line.*

Proof: Let l be a line of π_1, m a line of π_2. We have $l \nparallel m$ so that there exists exactly one point P of l which also lies on m. On the other hand let Q be any point of l. There is exactly one line $m' \parallel m$, hence one line of π_2 such that Q lies on m'. We have, therefore, a one-to-one correspondence between the points of l and the lines of π_2: the number of points of l is the same as the number of lines of π_2. Thus far we have shown: If two distinct pencils are given, then each line of one pencil contains as many points as there are lines in the other pencil. If π is any pencil, then it is certainly distinct from at least two of our three pencils, say $\pi \neq \pi_1$ and $\pi \neq \pi_2$. The number of points on a line of π_1 is equal to the number of lines in π and equal to the number of lines in π_2. The theorem follows easily.

AXIOM 3. There exist three distinct points A, B, C such that C does not lie on the line $A + B$. We also say that there exist three non-collinear points.

The lines $A + B$ and $A + C$ are not parallel: otherwise (containing the point A) they would be equal and C would lie on $A + B$. For the same reason $A + B \nparallel B + C$ and $A + C \nparallel B + C$. There do exist now at least three distinct pencils of parallel lines and Theorem 2.2 applies.

EXERCISE 1. Enumerate carefully (not being prejudiced at all about existence of points or lines or points on lines) all cases where the first two axioms hold but where axiom 3 is violated.

EXERCISE 2. Find the geometry with the *least* possible number of points in which all three axioms hold. Show that its structure is unique.

2. Dilatations and translations

In modern mathematics the investigation of the symmetries of a given mathematical structure has always yielded the most powerful results. Symmetries are maps which preserve certain properties. In our case they will be maps preserving "direction".

DEFINITION 2.3. A map σ of points into points is called a dilatation if it has the following property:

Let two distinct points P and Q and their images P' and Q' be given. If l' is the line $\parallel P + Q$ which passes through P', then Q' lies on l'.

We can give two examples.

1) Map all points onto one given point. We shall call this a degenerate dilatation and any other dilatation non-degenerate.

2) The identity map.

We can now immediately prove

THEOREM 2.3. *A dilatation σ is uniquely determined by the images P', Q' of two distinct points P and Q. Should $P' = Q'$, then σ is degenerate and all points are mapped into P'. Should $P' \neq Q'$, then σ is one-to-one and onto (every point is an image).*

Proof:

1) Let R be any point which is *not* on $P + Q$. Then $R + P \nparallel R + Q$. Let l' be the line through P' which is $\parallel R + P$. By definition of a dilatation R' must lie on l'. Similarly, R' will lie on the line l'' which is $\parallel R + Q$ and passes through Q'. The lines l' and l'' are not parallel and contain R'. We see that R' is uniquely determined.

2) Let now R be a point of $P + Q$, say $R \neq P$. Select any point S not on $P + Q$. The point R is not on $P + S$; otherwise R and P would lie on $P + Q$ and $P + S$, hence $P + Q = P + S$, contradicting the choice of S. The image S' of S is determined by 1); we know the

images P' and S' of P and S, and R is not on $P + S$. By 1) the image of R is determined.

3) Suppose $P' = Q'$. The degenerate map τ which maps every point into P' has the same effect as σ on P and Q. By the uniqueness which we have already shown $\sigma = \tau$.

4) Suppose $P' \neq Q'$ and let R' be a given point. We have to show that there exists a point R whose image is R'. Suppose first that R' does not lie on $P' + Q'$. Then $P' + R' \nparallel Q' + R'$. Let l_1 be the line $\parallel P' + R'$ which passes through P and l_2 the line $\parallel Q' + R'$ which passes through Q. Then $l_1 \nparallel l_2$. There is a point R which lies on both l_1 and l_2. If R would lie on $P + Q$, then $P + Q$ would meet l_1 in P and R and l_2 in Q and R so that $P + Q$ must equal one of the lines l_1 or l_2. But $P' + Q'$ is neither $\parallel$ to $P' + R'$ nor to $Q' + R'$. It follows now from 1) that R has R' as image. If R' lies on $P' + Q'$ we first select a point S' which is not on $P' + Q'$, find the point S whose image S' is, and argue with P' and S' instead of P' and Q'. This finishes the proof of the theorem.

As an immediate consequence we have the

Corollary. If the dilatation σ has two fixed points, then $\sigma = 1$, σ is the identity map.

Definition 2.4. Let σ be a non-degenerate dilatation, P a point. Any line containing P and σP shall be called a trace of P. If $P \neq \sigma P$, then the trace is unique and $= P + \sigma P$.

Theorem 2.4. *Let σ be a non-degenerate dilatation, P a point and l a trace of P. If Q lies on l, then σQ lies on l also.*

Proof: We can assume $Q \neq P$ so that $l = P + Q$. Then $\sigma Q \neq \sigma P$ and the line $\sigma P + \sigma Q \parallel l$ by definition of a dilatation. But l and $\sigma P + \sigma Q$ have the point σP in common and are, therefore, equal. This shows that σQ lies on l.

Theorem 2.4 has also an immediate

Corollary. The intersection of two non-parallel traces is a fixed point.

The possibilities for the traces of a non-singular dilatation σ are:

1) All lines, if and only if $\sigma = 1$.

2) All lines through a point P if $\sigma \neq 1$ and if P is a fixed point of σ. Indeed, any line through P is a trace and no other line could be a trace since $\sigma \neq 1$ can not have more than one fixed point.

3) A certain pencil of parallel lines if σ has no fixed point.

This exhausts all possibilities and suggests a definition.

DEFINITION 2.5. A non-singular dilatation τ shall be called a translation if either $\tau = 1$ or if τ has no fixed point. If τ is a translation $\neq 1$, then the traces of τ form a pencil π of parallel lines which we shall call the "direction" of τ.

THEOREM 2.5. *A translation τ is uniquely determined by the image of one point P.* (Notice that we do not claim the existence of translations with pre-assigned image point.)

Proof: Let l be a trace of τ which contains P. Any line parallel to l is also a trace of τ; let $l' \parallel l$, $l' \neq l$ and Q on l'. Then τQ must lie on l'. Put $m = P + Q$ and let m' be parallel to m and contain τP. By definition of a dilatation τQ must also lie on m'. τQ lies on both l' and m' and $l' \nparallel m'$ since $l \nparallel m$. Thus τQ is uniquely determined. But a dilation is uniquely determined by the image of two points.

We shall assume from now on that dilatations are non-singular unless we explicitly state the opposite. The definition of a dilatation can then be simplified to:

Whenever $P \neq Q$ we have $\sigma P \neq \sigma Q$ and $P + Q \parallel \sigma P + \sigma Q$.

Let σ_1, σ_2 be two dilatations. We can form the combined map $\sigma_1(\sigma_2(P))$ which we denote by $\sigma_1\sigma_2$. If $P \neq Q$, then $\sigma_2 P \neq \sigma_2 Q$ hence $\sigma_1\sigma_2 P \neq \sigma_1\sigma_2 Q$ and

$$P + Q \parallel \sigma_2 P + \sigma_2 Q \parallel \sigma_1\sigma_2 P + \sigma_1\sigma_2 Q.$$

The map $\sigma_1\sigma_2$ is again a dilatation.

Since a dilatation σ is one-to-one and onto we can form the inverse map which will also be one-to-one and onto. If $P \neq Q$ then $\sigma^{-1}P \neq \sigma^{-1}Q$ and $\sigma^{-1}P + \sigma^{-1}Q \parallel \sigma(\sigma^{-1}P) + \sigma(\sigma^{-1}Q) = P + Q$ which shows σ^{-1} to be a dilatation. The dilatations form, therefore, a group.

If τ is a translation, then τ^{-1} is also a translation. Suppose indeed that τ^{-1} has a fixed point P. Then $\tau^{-1}P = P$. Applying τ to it we get $P = \tau P$ which is only possible for the translation $\tau = 1$ in which case $\tau^{-1} = 1$ is also a translation.

Assume τ_1 and τ_2 are translations and suppose that $\tau_1\tau_2$ has a fixed point: $\tau_1\tau_2 P = P$. Then $\tau_2 P = \tau_1^{-1}P$. Thus τ_2 and τ_1^{-1} have the same effect on P which means that they are equal: $\tau_2 = \tau_1^{-1}$. Then $\tau_1\tau_2 = 1$ which is a translation and we see that $\tau_1\tau_2$ is a translation in any case. The translations form, therefore, a group.

Let σ be a dilatation and τ a translation. We wish to show that $\sigma\tau\sigma^{-1} = \tau_1$ is again a translation. If τ_1 has a fixed point P, then $\sigma\tau\sigma^{-1}P = P$ which implies $\tau\sigma^{-1}P = \sigma^{-1}P$, showing that $\sigma^{-1}P$ is a fixed point of τ. Hence $\tau = 1$, $\tau_1 = 1$ which is also a translation. Suppose now that $\tau \neq 1$ and that the pencil π is the direction of τ. $\sigma^{-1}P + \tau\sigma^{-1}P$ is the τ-trace of $\sigma^{-1}P$ and is, therefore, a line of π. Since σ is a dilatation, $\sigma^{-1}P + \tau\sigma^{-1}P \parallel \sigma(\sigma^{-1}P) + \sigma(\tau\sigma^{-1}P) = P + \sigma\tau\sigma^{-1}P$. The line $P + \sigma\tau\sigma^{-1}P$ is also in π and is a $\sigma\tau\sigma^{-1}$-trace of P. This shows that $\sigma\tau\sigma^{-1}$ has also the direction π. Let us gather all these results:

THEOREM 2.6. *The dilatations form a group D and the translations form an invariant sub-group T of D. If σ is a dilatation and $\tau \neq 1$ a translation, then τ and $\sigma\tau\sigma^{-1}$ have the same direction.*

THEOREM 2.7. *Identity and the translations having a given pencil π of parallel lines as direction form a group.*

Proof: If $\tau \neq 1$, then $P + \tau P = \tau P + \tau^{-1}(\tau P)$ is a τ-trace of P and a τ^{-1}-trace of τP.

If $\tau_1 \neq 1$ and $\tau_2 \neq 1$ have the direction π, then $P + \tau_2 P$ lies in π and contains $\tau_2 P$. It contains, therefore, also $\tau_1\tau_2 P$, by Theorem 2.4. If $\tau_1\tau_2 P = P$, then $\tau_1\tau_2 = 1$; if $\tau_1\tau_2 P \neq P$, then $P + \tau_2 P = P + \tau_1\tau_2 P$ which is a trace of $\tau_1\tau_2$.

THEOREM 2.8. *If translations with different directions exist, then T is a commutative group.*

Proof:

1) Suppose τ_1 and τ_2 have different directions. By Theorem 2.6 the translation $\tau_1\tau_2\tau_1^{-1}$ has the same direction as τ_2 , hence the same direction as τ_2^{-1}. If $\tau_1\tau_2\tau_1^{-1}\tau_2^{-1} \neq 1$, then $\tau_1\tau_2\tau_1^{-1}\tau_2^{-1}$ has the same direction as τ_2 . But τ_1 and $\tau_2\tau_1^{-1}\tau_2^{-1}$ have the same direction as τ_1 so that $\tau_1\tau_2\tau_1^{-1}\tau_2^{-1}$ has also the direction τ_1 . This is a contradiction and consequently $\tau_1\tau_2\tau_1^{-1}\tau_2^{-1} = 1$ which shows $\tau_1\tau_2 = \tau_2\tau_1$.

2) Suppose τ_1 and τ_2 have the same direction. By assumption there exists a translation τ_3 whose direction is different from that of τ_1 and τ_2 . Therefore $\tau_3\tau_1 = \tau_1\tau_3$. The translation $\tau_2\tau_3$ must also have a direction different from that of τ_1 ; otherwise $\tau_2^{-1}\cdot\tau_2\tau_3 = \tau_3$ would have the same direction. We know, therefore, also that

$$\tau_1\cdot(\tau_2\tau_3) = (\tau_2\tau_3)\cdot\tau_1 = \tau_2(\tau_3\tau_1) = \tau_2(\tau_1\tau_3).$$

From $\tau_1\tau_2\tau_3 = \tau_2\tau_1\tau_3$ we get now $\tau_1\tau_2 = \tau_2\tau_1$.

REMARK. It is conceivable that the geometry contains only translations with a single direction π (aside from identity). It is an undecided question whether T has to be commutative in this case. Probably there do exist counter-examples, but none is known.

3. Construction of the field

We can hope for a "good" geometry only if the geometry has enough symmetries. We postulate, therefore,

AXIOM 4a. Given any two points P and Q, there exists a translation τ_{PQ} which moves P into Q:

$$\tau_{PQ}(P) = Q.$$

REMARK. Since the image of one point determines a translation, τ_{PQ} is unique; especially $\tau_{PP} = 1$. Since we now obviously have translations with different directions, our group T of translations will be commutative. The geometric meaning of axiom 4a is obvious, another geometric interpretation shall be given later.

We shall define new maps which will send T into T.

DEFINITION 2.6. A map $\alpha : T \rightarrow T$ shall be called a "trace preserving homomorphism" if:

1) It is a homomorphism of T, meaning by this that

$$(\tau_1\tau_2)^\alpha = \tau_1^\alpha\tau_2^\alpha .$$

(We use the symbol τ^α for the image of τ under the map α for convenience.)

2) It preserves traces, or, more precisely: the traces of τ are among the traces of τ^α. This means that either $\tau^\alpha = 1$, or that τ and τ^α have the same direction.

Important examples of trace-preserving homomorphisms are the following.

a) Map every translation τ of T onto identity: $\tau^\alpha = 1$ for all $\tau \varepsilon T$. Clearly both conditions are satisfied. We denote this very special map by 0 since $\tau^0 = 1$ is very suggestive.

b) The identity which will be denoted by 1. Thus $\tau^1 = \tau$ for all $\tau \varepsilon T$.

c) Map each τ onto its inverse τ^{-1}. We denote this map of course by τ^{-1}. We have

$$(\tau_1\tau_2)^{-1} = \tau_2^{-1}\tau_1^{-1} = \tau_1^{-1}\tau_2^{-1}$$

since T is commutative; τ^{-1} and τ have the same traces.

d) Let σ be a fixed dilatation. Map each τ onto $\sigma\tau\sigma^{-1}$. We know already that τ and $\sigma\tau\sigma^{-1}$ have the same traces. Furthermore

$$\sigma(\tau_1\tau_2)\sigma^{-1} = \sigma\tau_1\sigma^{-1}\cdot\sigma\tau_2\sigma^{-1}$$

which shows condition 1. We do not give a special name to this map.

The set of *all* trace-preserving homomorphisms shall be denoted by k.

DEFINITION 2.7. Let α, β be elements of k. We may construct a new map which sends τ into $\tau^{\alpha}\cdot\tau^{\beta}$ and shall call this map $\alpha + \beta$. Thus

$$\tau^{\alpha+\beta} = \tau^{\alpha}\cdot\tau^{\beta}.$$

We may also send τ into $(\tau^{\beta})^{\alpha}$ and call this map $\alpha\cdot\beta$. Therefore

$$\tau^{\alpha\beta} = (\tau^{\beta})^{\alpha}.$$

THEOREM 2.9. *If α and β belong to k, then $\alpha + \beta$ and $\alpha\beta$ belong to k. Under this definition the set k becomes an associative ring with unit element* 1.

Proof:

1) $(\tau_1\tau_2)^{\alpha+\beta} = (\tau_1\tau_2)^{\alpha}(\tau_1\tau_2)^{\beta}$ (by definition of $\alpha + \beta$)

$= \tau_1^{\alpha}\tau_2^{\alpha}\tau_1^{\beta}\tau_2^{\beta}$ (since α, $\beta \in k$)

$= \tau_1^{\alpha}\tau_1^{\beta}\tau_2^{\alpha}\tau_2^{\beta}$ (since T is commutative).

Hence

$$(\tau_1\tau_2)^{\alpha+\beta} = \tau_1^{\alpha+\beta}\tau_2^{\alpha+\beta}.$$

If $\tau = 1$, then $\tau^{\alpha} = \tau^{\beta} = 1$, hence $\tau^{\alpha+\beta} = 1$.

If $\tau \neq 1$, then its direction π occurs among the traces of τ^{α} and τ^{β} and, therefore, also among the traces of $\tau^{\alpha}\cdot\tau^{\beta} = \tau^{\alpha+\beta}$. This shows $\alpha + \beta \in k$.

2) $(\tau_1\tau_2)^{\alpha\beta} = ((\tau_1\tau_2)^{\beta})^{\alpha} = (\tau_1^{\beta}\tau_2^{\beta})^{\alpha} = (\tau_1^{\beta})^{\alpha}(\tau_2^{\beta})^{\alpha}$

$= \tau_1^{\alpha\beta}\tau_2^{\alpha\beta}.$

The traces of τ are among those of τ^{β}, hence among those of $(\tau^{\beta})^{\alpha} = \tau^{\alpha\beta}$.

3) If $\alpha, \beta, \gamma \in k$, we show $(\alpha + \beta) + \gamma = \alpha + (\beta + \gamma)$:

$$\tau^{(\alpha+\beta)+\gamma} = \tau^{\alpha+\beta}\tau^{\gamma} = \tau^{\alpha}\tau^{\beta}\tau^{\gamma} = \tau^{\alpha}\tau^{\beta+\gamma} = \tau^{\alpha+(\beta+\gamma)}$$

(we used the associativity of T).

4) $$\alpha + \beta = \beta + \alpha:$$

$$\tau^{\alpha+\beta} = \tau^{\alpha}\tau^{\beta} = \tau^{\beta}\tau^{\alpha} \qquad \text{(since } T \text{ is commutative)}$$
$$= \tau^{\beta+\alpha}.$$

5) $$0 + \alpha = \alpha:$$

$$\tau^{0+\alpha} = \tau^{0}\cdot\tau^{\alpha} = 1\cdot\tau^{\alpha} = \tau^{\alpha}.$$

6) $$\alpha + (-1)\alpha = 0:$$

$$\tau^{\alpha+(-1)\alpha} = \tau^{\alpha}\cdot\tau^{(-1)\alpha} = \tau^{\alpha}\cdot(\tau^{\alpha})^{-1} = 1 = \tau^{0}.$$

This shows that k is a commutative group under addition.

7) $$(\beta + \gamma)\alpha = \beta\alpha + \gamma\alpha:$$

$$\tau^{(\beta+\gamma)\alpha} = (\tau^{\alpha})^{\beta+\gamma} = (\tau^{\alpha})^{\beta}(\tau^{\alpha})^{\gamma} = \tau^{\beta\alpha}\cdot\tau^{\gamma\alpha}$$
$$= \tau^{\beta\alpha+\gamma\alpha}.$$

Notice that only definitions of addition and multiplication are used.

8) $$\alpha(\beta + \gamma) = \alpha\beta + \alpha\gamma:$$

$$\tau^{\alpha(\beta+\gamma)} = (\tau^{\beta+\gamma})^{\alpha} = (\tau^{\beta}\tau^{\gamma})^{\alpha} = (\tau^{\beta})^{\alpha}(\tau^{\gamma})^{\alpha}$$
$$= \tau^{\alpha\beta}\tau^{\alpha\gamma} = \tau^{\alpha\beta+\alpha\gamma}.$$

Here we had to use that α is a homomorphism.

9) $$(\alpha\beta)\gamma = \alpha(\beta\gamma):$$

$$\tau^{(\alpha\beta)\gamma} = (\tau^{\gamma})^{\alpha\beta} = ((\tau^{\gamma})^{\beta})^{\alpha} = (\tau^{\beta\gamma})^{\alpha} = \tau^{\alpha(\beta\gamma)}.$$

10) $$1\cdot\alpha = \alpha \quad\text{and}\quad \alpha\cdot 1 = \alpha:$$

$$\tau^{1\cdot\alpha} = (\tau^{\alpha})^{1} = \tau^{\alpha} \quad\text{and}\quad \tau^{\alpha\cdot 1} = (\tau^{1})^{\alpha} = \tau^{\alpha}.$$

This establishes all axioms for a ring.

In our next theorem we use axiom 4a to the full extent.

THEOREM 2.10. *Let $\alpha \in k$, $\alpha \neq 0$ and P be a given point. There exists a unique dilatation σ which has P as fixed point and such that*

$$\tau^\alpha = \sigma\tau\sigma^{-1} \quad \text{for all} \quad \tau \in T.$$

Proof:

1) Suppose such a σ exists. Let Q be any point and select for τ the translation τ_{PQ}. Then

$$\tau_{PQ}^\alpha = \sigma\tau_{PQ}\sigma^{-1}.$$

Apply this to the point P remembering that σ leaves P fixed:

$$\tau_{PQ}^\alpha(P) = \sigma\tau_{PQ}\sigma^{-1}(P) = \sigma\tau_{PQ}(P) = \sigma(Q).$$

Thus

$$(2.1) \qquad \sigma(Q) = \tau_{PQ}^\alpha(P).$$

This shows how to compute the image of a given point Q under σ: One takes τ_{PQ}, computes τ_{PQ}^α and applies this to P. This proves not only the uniqueness but shows us also which map σ has to be tried in order to prove our theorem.

2) Let now $\alpha \in k$ and allow even $\alpha = 0$. Define a map σ on the points Q by formula (2.1). We contend that σ is a dilatation, but possibly degenerate.

Let Q and R be two distinct points. The translation $\tau_{QR}\tau_{PQ}$ moves P into R since $\tau_{QR}\tau_{PQ}(P) = \tau_{QR}(Q) = R$. Therefore

$$\tau_{QR}\tau_{PQ} = \tau_{PR}.$$

If we apply α to this equation, we get

$$\tau_{QR}^\alpha\tau_{PQ}^\alpha = \tau_{PR}^\alpha.$$

Applying both sides to the point P we obtain

$$\tau_{QR}^\alpha(\tau_{PQ}^\alpha(P)) = \tau_{PR}^\alpha(P),$$

hence, by formula (2.1),

$$(2.2) \qquad \tau_{QR}^\alpha(\sigma(Q)) = \sigma(R).$$

Let l be a line through $\sigma(Q)$ which is parallel to $Q + R$; l is a trace of τ_{QR} and hence a trace of τ_{QR}^α. $\sigma(Q)$ lies on this trace of τ_{QR}^α; thus its image $\sigma(R)$ must also lie on l. But this is the condition σ must satisfy to be a dilatation. By (2.1) we have $\sigma(P) = \tau_{PP}^\alpha(P) = 1^\alpha(P) = 1(P) = P$ so that σ has P as fixed point. If σ is degenerate, then every point is mapped into P and (2.1) gives $P = \tau_{PQ}^\alpha(P)$; hence $\tau_{PQ}^\alpha = 1$ for all Q. Any translation τ has the form τ_{PQ}, namely for

$Q = \tau(P)$. Therefore $\tau^\alpha = 1$ for all $\tau \varepsilon T$ which means $\alpha = 0$. If $\alpha \neq 0$, then σ is not degenerate.

Let now $\alpha \neq 0$. Since we know $\sigma(P) = P$, we can write (2.1) in the form

$$\sigma(Q) = \tau_{PQ}^\alpha \sigma(P)$$

or

$$Q = \sigma^{-1}\tau_{PQ}^\alpha \sigma(P).$$

This means that the translation $\sigma^{-1}(\tau_{PQ}^\alpha)(\sigma^{-1})^{-1}$ moves P into Q and is consequently τ_{PQ}. Hence

$$\sigma^{-1}\tau_{PQ}^\alpha\sigma = \tau_{PQ}$$

or

$$\tau_{PQ}^\alpha = \sigma\tau_{PQ}\sigma^{-1}.$$

We have already mentioned that τ_{PQ} can be any translation τ, and have, therefore,

$$\tau^\alpha = \sigma\tau\sigma^{-1} \quad \text{for all} \quad \tau \varepsilon T.$$

But this is the contention of our theorem.

REMARK. Example d) of the trace-preserving homomorphisms shows the converse. If σ is a dilatation, then $\tau \rightarrow \sigma\tau\sigma^{-1}$ is a trace-preserving homomorphism α; it can not be 0 since $\sigma\tau\sigma^{-1} = 1$ holds only for $\tau = 1$ and not for all τ. We see, therefore, that example d) has already exhausted all possibilities for a trace-preserving homomorphism. To a given $\alpha \neq 0$ we can find a σ and even prescribe the fixed point for σ.

THEOREM 2.11. *k is a field (possibly not commutative under multiplication).*

Proof: Let $\alpha \varepsilon k$ and $\alpha \neq 0$. Find a dilatation σ such that $\tau^\alpha = \sigma\tau\sigma^{-1}$. The map which sends τ into $\sigma^{-1}\tau\sigma = \sigma^{-1}\tau(\sigma^{-1})^{-1}$ is also an element of k which we shall call α^{-1}. Thus

$$\tau^\alpha = \sigma\tau\sigma^{-1} \quad \text{and} \quad \tau^{\alpha^{-1}} = \sigma^{-1}\tau\sigma.$$

Now,

$$\tau^{\alpha\cdot\alpha^{-1}} = (\tau^{\alpha^{-1}})^\alpha = \sigma(\sigma^{-1}\tau\sigma)\sigma^{-1} = \tau = \tau^1$$

and

$$\tau^{\alpha^{-1}\alpha} = (\tau^{\alpha})^{\alpha^{-1}} = \sigma^{-1}(\sigma\tau\sigma^{-1})\sigma = \tau = \tau^{1}$$

which shows $\alpha\alpha^{-1} = \alpha^{-1}\alpha = 1$ and establishes the existence of an inverse (its uniqueness follows from group theory, provided the existence is established).

THEOREM 2.12. *If* $\tau^{\alpha} = 1$ *holds for one* α *and one* τ, *then either* $\alpha = 0$ *or* $\tau = 1$. *If* $\tau^{\alpha} = \tau^{\beta}$ *holds for particular* α, β, τ, *then either* $\tau = 1$ *or* $\alpha = \beta$.

Proof: Suppose $\tau^{\alpha} = 1$ and $\alpha \neq 0$. Apply α^{-1} :

$$(\tau^{\alpha})^{\alpha^{-1}} = 1^{\alpha^{-1}} = 1 \quad \text{or} \quad \tau = 1.$$

Suppose now $\tau^{\alpha} = \tau^{\beta}$. Multiplying by $\tau^{-\beta}$ we have $\tau^{\alpha-\beta} = 1$. If $\alpha \neq \beta$, then $\tau = 1$.

(The element $-\beta$ is of course $(-1)\beta$ and also $\beta(-1)$, since $\beta + \beta\cdot(-1) = \beta(1 + (-1)) = \beta\cdot 0 = 0$. The rule $\beta \cdot 0 = 0$ holds in every distributive ring but is also checked directly: $\tau^{\beta\cdot 0} = (\tau^{0})^{\beta} = 1^{\beta} = 1$.)

4. *Introduction of coordinates*

Our geometry is still not "symmetric" enough. We shall need one more axiom which we shall put into two forms; the equivalence of the two forms will be a separate theorem.

AXIOM 4b. If τ_1 and τ_2 are translations with the same traces and if $\tau_1 \neq 1$, $\tau_2 \neq 1$, $\tau_1 \neq \tau_2$, then there exists an α in k such that $\tau_2 = \tau_1^{\alpha}$.

REMARK. If $\tau_2 = \tau_1$, then $\tau_2 = \tau_1^{1}$; if $\tau_2 = 1$, then $\tau_2 = \tau_1^{0}$. Thus only $\tau_1 \neq 1$ is needed, but we wanted to state the axiom in its weakest form. Theorem 2.12 shows that the α is unique.

Now we state the other form of the axiom.

AXIOM 4b P (for a given point P). Given two points Q and R such that P, Q, R are distinct but lie on a line; then there exists a dilatation σ which has P as fixed point and moves Q into R.

THEOREM 2.13. *If axiom* 4b P *holds for one point* P, *then axiom* 4b *is true, and axiom* 4b *implies axiom* 4b P *for all points* P.

Proof:

1) Assume axiom 4b P true for a particular P. Let τ_1 and τ_2 be translations satisfying the hypothesis of axiom 4b and set $\tau_1 P = Q$ and $\tau_2 P = R$. Since τ_1 and τ_2 have the same traces, the points P, Q, R are collinear and are distinct by the assumptions on τ_1 and τ_2. Let σ be the dilatation with fixed point P such that $\sigma(Q) = R$. $\sigma\tau_1\sigma^{-1}$ is a translation and $\sigma\tau_1\sigma^{-1}(P) = \sigma\tau_1(P) = \sigma(Q) = R$. Hence $\sigma\tau_1\sigma^{-1} = \tau_2$. Let $\alpha \in k$ such that $\tau^\alpha = \sigma\tau\sigma^{-1}$. Then $\tau_2 = \tau_1^\alpha$.

2) Assume that axiom 4b is true and let P, Q, R be three distinct collinear points. Put $\tau_1 = \tau_{PQ}$, $\tau_2 = \tau_{PR}$. Since P, Q, R are collinear, τ_1 and τ_2 have the same traces, $\tau_1 \neq 1$, $\tau_2 \neq 1$ and $\tau_1 \neq \tau_2$. By axiom 4b there exists an $\alpha \neq 0$ such that $\tau_2 = \tau_1^\alpha$. By Theorem 2.10 there exists a dilatation with fixed point P such that $\tau^\alpha = \sigma\tau\sigma^{-1}$ for all $\tau \in T$, especially

$$\tau_2 = \sigma\tau_1\sigma^{-1}, \qquad \tau_2\sigma = \sigma\tau_1 .$$

If we apply this to P we get $\tau_2(P) = \sigma\tau_1(P)$ or $R = \sigma(Q)$; i.e., σ is the desired dilatation.

The geometric meaning of axiom 4b P is clear and another one shall be given later.

THEOREM 2.14. *Let $\tau_1 \neq 1$, $\tau_2 \neq 1$ be translations with different directions. To any translation $\tau \in T$ there exist unique elements α, β in k such that*

$$\tau = \tau_1^\alpha \tau_2^\beta = \tau_2^\beta \tau_1^\alpha .$$

Proof: The commutativity is clear since T is commutative.

1) Let P be any point and suppose $\tau(P) = Q$. Let l_1 be a τ_1-trace through P and l_2 a τ_2-trace through Q. Then $l_1 \nparallel l_2$ so that there is a point R on l_1 and l_2.

The translation τ_{PR} is either 1 or has the same direction as τ_1. The translation τ_{RQ} is also either 1 or has the same direction as τ_2. By axiom 4b and the remark there do exist elements α and β in k such that $\tau_{PR} = \tau_1^\alpha$ and $\tau_{RQ} = \tau_2^\beta$. Then $\tau_2^\beta\tau_1^\alpha(P) = \tau_{RQ}\tau_{PR}(P) = Q$ and hence $\tau_{PQ} = \tau_2^\beta\tau_1^\alpha$. But τ_{PQ} is our τ.

2) If $\tau_1^\alpha\tau_2^\beta = \tau_1^\gamma\tau_2^\delta$, then $\tau_1^{\alpha-\gamma} = \tau_2^{\beta-\delta}$. If $\tau_1^{\alpha-\gamma} \neq 1$, then the left side of $\tau_1^{\alpha-\gamma} = \tau_2^{\delta-\beta}$ has the direction of τ_1, the right side the direction of τ_2 which is not possible. Thus $\tau_1^{\alpha-\gamma} = 1$ and $\tau_2^{\delta-\beta} = 1$. Theorem 2.12 shows $\alpha = \gamma$ and $\beta = \delta$.

We are now ready to introduce coordinates.

In ordinary geometry one describes them by selecting an origin O, by drawing two coordinate axis and by marking "unit points" on each of the axis.

We shall do precisely the same. We select a point O as "origin" and two translations $\tau_1 \neq 1$ and $\tau_2 \neq 1$ with different traces. The naive meaning of τ_1 and τ_2 is the following: The τ_1-trace through O and the τ_2-trace through O shall be thought of as the coordinate axis and the two points $\tau_1(O)$ and $\tau_2(O)$ as the "unit points".

Let now P be any point. Write the translation τ_{OP} in the form

$$\tau_{OP} = \tau_1^{\xi}\tau_2^{\eta}$$

with the unique $\xi, \eta \varepsilon k$ and assign to P the pair (ξ, η) as coordinates. If conversely (ξ, η) is a given pair, then $\tau_1^{\xi}\tau_2^{\eta}$ is a translation, it takes O into a certain point P so that $\tau_{OP} = \tau_1^{\xi}\tau_2^{\eta}$, and this point P will have the given coordinates (ξ, η). We write $P = (\xi, \eta)$.

For $P = O$ we get $\tau_{OO} = 1 = \tau_1^0\tau_2^0$ so that the origin has coordinates $(0, 0)$.

For $(1, 0)$ we have $\tau_1^1\tau_2^0 = \tau_1 = \tau_{OP}$ so that $\tau_1(O)$ is indeed the "unit point" on the first axis and similarly $\tau_2(O)$ the unit point on the second axis.

Let now l be any line, $P = (\alpha, \beta)$ a point on l and $\tau = \tau_1^{\gamma}\tau_2^{\delta}$ a translation $\neq 1$ which has l as trace. Let $Q = (\xi, \eta)$ be any point on l. Then τ_{PQ} is either 1 or has the direction of τ. By axiom 4b and the remark we have

$$\tau_{PQ} = \tau^t$$

with some $t \varepsilon k$.

Conversely, for any t, l will appear among the traces of τ^t; $\tau^t(P) = Q$ is some point on l which shows that every τ^t is some τ_{PQ} with Q on l. We also know that t is uniquely determined by τ_{PQ} (Theorem 2.12).

To get the coordinates of Q we must compute τ_{OQ}. We have

$$\tau_{OQ} = \tau_{PQ}\tau_{OP} = (\tau_1^{\gamma}\tau_2^{\delta})^t\tau_1^{\alpha}\tau_2^{\beta} = \tau_1^{t\gamma+\alpha}\tau_2^{t\delta+\beta}.$$

Thus $Q = (t\gamma + \alpha, t\delta + \beta)$ will range over the points of l if t ranges over k. How much freedom can we have in the choices of $\alpha, \beta, \gamma, \delta$ and still be certain that $Q = (t\gamma + \alpha, t\delta + \beta)$ describes the points of a line?

Since $P = \tau_1^{\alpha}\tau_2^{\beta}$ can be any point, α, β are arbitrary; $\tau = \tau_1^{\gamma}\tau_2^{\delta}$ had to be a translation $\neq 1$ which means that γ and δ are not both 0. If this is the case, then the line through P which is a τ-trace is the line that our formula describes.

We may abbreviate in vectorial notation:
We put $A = (\gamma, \delta)$ and $P = (\alpha, \beta)$. Then

$$Q = P + tA,$$

t ranging over k, gives the points of a line. $A \neq 0$ is the only restriction.

The reader may be more familiar with another form of the equation of a straight line:

Put $Q = (x, y)$. Then $x = t\gamma + \alpha$, $y = t\delta + \beta$.

We consider two cases:

a) $\gamma \neq 0$. Then $t = x\gamma^{-1} - \alpha\gamma^{-1}$ and consequently $y = x\gamma^{-1}\delta + \beta - \alpha\gamma^{-1}\delta$. This is the form

$$y = xm + b.$$

Converse, if $y = xm + b$, put $x = t$, $y = tm + b$ and we have a "parametric form" with $\gamma = 1$, $\delta = m$, $\alpha = 0$, $\beta = b$.

b) $\gamma = 0$. Then $\delta \neq 0$ so that y is arbitrary and $x = \alpha$.

We have certainly recovered the usual description of points and lines by coordinates.

5. *Affine geometry based on a given field*

We shall start now at the other end. Suppose that k is a given field and let us build up an affine geometry based on k.

We define a point P to be an ordered pair (ξ, η) of elements $\xi, \eta \in k$. We shall identify P also with the vector (ξ, η) and use vector notation. A line l will be defined as the point set $P + tA$ where P is a given point, A a given non-zero vector and where t ranges over k. The relation that a point Q lies on a line l shall merely mean that Q is an element of the point set we called l. Our problem is now to prove that this geometry satisfies all our axioms.

Let us first consider the possible intersections of a line $P + tA$ and $Q + uB$ where A and B are non-zero vectors and where t and u range over k. For a common point we must have

$$P + tA = Q + uB$$

and we must find the t and u satisfying this equation which we shall now write in the form

$$tA - uB = Q - P. \tag{2.3}$$

Case 1. A and B are left linearly independent—an equation $xA + yB = 0$ always implies that $x = y = 0$. Then any vector can be expressed uniquely in the form $xA + yB$ and (2.3) merely requires that $Q - P$ be expressed in this form. We see that there is exactly one t and u solving (2.3) and this means that the intersection is exactly one point.

Case 2. A and B are left linearly dependent—elements α, β in k exist which are not both zero such that $\alpha A + \beta B = 0$. Then $\alpha \neq 0$; for, otherwise $\beta B = 0$, hence $\beta = 0$, since we have assumed $B \neq 0$. Similarly $\beta \neq 0$. Then $B = \gamma A$ where $\gamma = -\beta^{-1}\alpha \neq 0$.

We can then simplify the line $Q + uB = Q + u\gamma A = Q + vA$ with $v = u\gamma$; if u ranges over k, so will v. We may assume, therefore, $B = A$ and (2.3) becomes

$$(t - u)A = Q - P. \tag{2.4}$$

If $Q - P$ is not a left multiple of A, we can not possibly solve this equation—the two lines do not intersect, they are parallel. If $Q - P = \delta A$ is a left multiple of A, then $Q = P + \delta A$ and the line $Q + uA$ becomes $P + (u + \delta)A$. If u ranges over k, so does $u + \delta$ and we see that the two lines are equal and, therefore, parallel again.

We see now: two lines $P + tA$ and $Q + uB$ are parallel if and only if $B = \gamma A$. A pencil of parallel lines consists of the lines $P + tA$ for a fixed A if we let P range over all points.

Our analysis has also shown that there can be at most one line containing two distinct points. Let now $P \neq Q$ be two distinct points. Then $Q - P \neq 0$ so that $P + t(Q - P)$ is a line. For $t = 0$ we get P, for $t = 1$ we get Q as points on this line. Thus axiom 1 is verified.

If $P + tA$ is a given line and Q a given point, then $Q + uA$ is a parallel line and contains Q (for $u = 0$). Our discussion shows the uniqueness. This is axiom 2.

The points $A = (0, 0)$, $B = (1, 0)$, $C = (0, 1)$ are not collinear: the line $A + t(B - A) = (t, 0)$ contains A and B but not C. Axiom 3 holds.

Consider next the following map σ: select $\alpha \in k$ and any vector C.

Put

$$\sigma X = \alpha X + C. \tag{2.5}$$

If $\alpha = 0$, then every point is mapped onto C and σ is a dilatation.

Let $\alpha \neq 0$. The image of the line $P + tA$ is $\alpha(P + tA) + C = \alpha P + C + \alpha tA = (\alpha P + C) + uA$ since $u = \alpha t$ will range over k if t does.

The image of the line $P + tA$ is the line $(\alpha P + C) + uA$ which is parallel to $P + tA$. Thus σ is a dilatation and a non-degenerate one, since a line goes into a line.

The fixed points of σ must satisfy

$$X = \alpha X + C, \qquad (1 - \alpha)X = C.$$

We have two cases:

1) $\alpha \neq 1$; then $(1 - \alpha)^{-1}C$ is the only fixed point.

2) $\alpha = 1$; if $C \neq 0$, there is no fixed point, if $C = 0$, then $\sigma = 1$, the identity.

We see that the translations among our maps are of the form $X \to X + C$. Let us denote this translation by τ_C. Obviously $\tau_{Q-P}(P) = P + (Q - P) = Q$ which verifies axiom 4a and shows that the translations τ_C are all translations of our geometry.

To verify axiom 4b P let P, Q, R be three collinear points, $P \neq Q$, $P \neq R$. Since R lies on the line $P + t(Q - P)$ which contains P and Q, we can determine α in such a way that

$$P + \alpha(Q - P) = R.$$

This α will be $\neq 0$ since $R \neq P$. With this α we form the dilatation

$$\sigma(X) = \alpha X + P - \alpha P.$$

Clearly $\sigma(P) = P$ and $\sigma(Q) = R$ which shows that axiom 4b P holds for all P and also that any dilatation with fixed point P is of this form. We know, therefore, that (2.5) gives all dilatations of our geometry.

Next we determine all trace-preserving homomorphisms of the translation group T. By Theorem 2.10 we may proceed as follows.

We select as the point P the origin $(0, 0)$. If $\sigma(X) = \alpha X + C$ has $(0, 0)$ as fixed point then $(0, 0) = C$. The dilatations with P as fixed point are of the form

$$\sigma X = \alpha X, \qquad \alpha \neq 0.$$

The α is uniquely determined by σ. The maps $\tau \to \sigma\tau\sigma^{-1}$ are then all trace-preserving homomorphisms $\neq 0$ and different dilatations σ will give different homomorphisms. We have $\sigma^{-1}X = \alpha^{-1}X$. For the translation τ_C we get as image

$$\sigma\tau_C\sigma^{-1}(X) = \sigma\tau_C(\alpha^{-1}X) = \sigma(\alpha^{-1}X + C)$$
$$= \alpha(\alpha^{-1}X + C) = X + \alpha C = \tau_{\alpha C} .$$

We see that for a given $\alpha \neq 0$ the map $\tau_C \to \tau_{\alpha C}$ is the desired trace-preserving homomorphism. To it we have to add the 0-map which maps each τ_C into $1 = \tau_0$. This also can be written in the form $\tau_C \to \tau_{\alpha C}$, namely for $\alpha = 0$.

Denote now the map $\tau_C \to \tau_{\alpha C}$ for $\alpha \in k$ by $\bar{\alpha}$ and let $\bar{k}$ be the field of all trace-preserving homomorphisms. We have just seen that $\alpha \to \bar{\alpha}$ is a one-to-one correspondence between the field k and the field $\bar{k}$. We contend that it is an isomorphism and have to this effect merely to show that $\overline{\alpha + \beta} = \bar{\alpha} + \bar{\beta}$ and that $\overline{\alpha\beta} = \bar{\alpha}\bar{\beta}$.

The map $\bar{\alpha}$ sends τ_C into $\tau_{\alpha C}$, $\bar{\beta}$ sends it into $\tau_{\beta C}$ and $\bar{\alpha} + \bar{\beta}$ (by definition of addition in $\bar{k}$) into $\tau_{\alpha C} \cdot \tau_{\beta C}$.
But

$$\tau_{\alpha C}\tau_{\beta C}(X) = \tau_{\alpha C}(X + \beta C) = X + \beta C + \alpha C$$
$$= X + (\alpha + \beta)C = \tau_{(\alpha+\beta)C}(X).$$

$\bar{\alpha} + \bar{\beta}$ sends, therefore τ_C into $\tau_{(\alpha+\beta)C}$ and $\overline{\alpha + \beta}$ does the same.

The map $\bar{\beta}$ sends τ_C into $\tau_{\beta C}$ and $\bar{\alpha}\bar{\beta}$ sends it into the image of $\tau_{\beta C}$ under $\bar{\alpha}$ which is $\tau_{\alpha\beta C}$. $\overline{\alpha\beta}$ does the same. The fields k and $\bar{k}$ are indeed isomorphic under the map $\alpha \to \bar{\alpha}$.

Now we introduce coordinates based on the field $\bar{k}$. We select our origin (0, 0) also as origin O for the coordinates in $\bar{k}$, select $\tau_1 = \tau_C$, $\tau_2 = \tau_D$ where $C = (1, 0)$ and $D = (0, 1)$. Let (ξ, η) be a given point P. Then

$$\tau_1^{\bar{\xi}} = \tau_C^{\bar{\xi}} = \tau_{\xi C} \quad \text{and} \quad \tau_2^{\bar{\eta}} = \tau_D^{\bar{\eta}} = \tau_{\eta D} ;$$

$$\tau_{\xi C}\tau_{\eta D}(0, 0) = \tau_{\xi C}(\eta D) = \eta D + \xi C = (\xi, \eta) = P.$$

The translation $\tau_1^{\bar{\xi}}\tau_2^{\bar{\eta}}$ will send the origin into P. The prescription was to assign the coordinates $(\bar{\xi}, \bar{\eta})$ to the point P. We see, therefore, that $P = (\xi, \eta)$ receives merely the coordinates $(\bar{\xi}, \bar{\eta})$ if we base the coordinates on $\bar{k}$.

This means that the coordinates are "the same" apart from our isomorphism and shows that on the whole we have recovered our original field k. Notice that the isomorphism $k \leftrightarrow \bar{k}$ is canonical.

6. *Desargues' theorem*

Assume now of our geometry only that the first three axioms hold. The following theorem may or may not hold in our geometry (see Figure 1).

THEOREM 2.15. *Let l_1, l_2, l_3 be distinct lines which are either parallel or meet in a point P. Let Q, Q' be points on l_1, R, R' points on l_2 and S, S' points on l_3 which are distinct from P if our lines meet. We assume*

$$Q + R \parallel Q' + R' \quad \textit{and} \quad Q + S \parallel Q' + S'$$

and contend that then

$$R + S \parallel R' + S'.$$

This theorem is called Desargues' theorem.

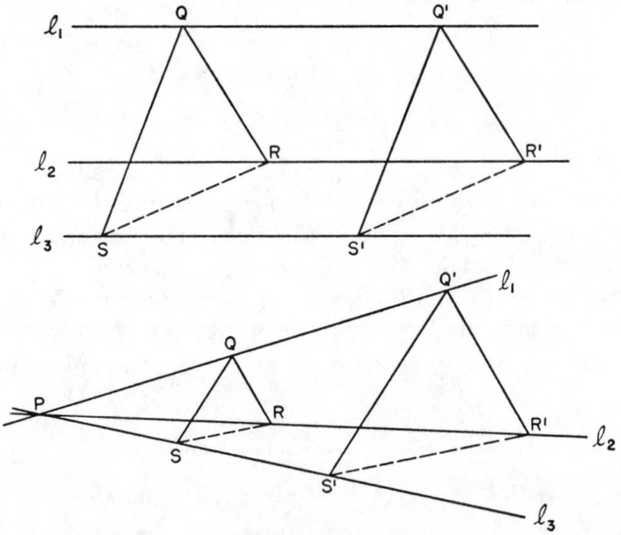

Figure 1

REMARK. There are certain trivial cases in which this theorem is obviously true:

1) If $Q = Q'$, then $Q + R = Q' + R'$ and since l_2 and $Q + R$ have R in common but not Q it follows that $R = R'$ and $S = S'$. Thus we may assume, if we wish, that the points are distinct.

2) If Q, R, S are collinear, then Q', R', S' are collinear and the statement is true again.

If the lines l_1, l_2, l_3 are parallel we shall call our theorem $\mathbf{D}_a$. If they meet in a point P, then we shall call it **DP**.

THEOREM 2.16. *Axiom* 4a *implies* $\mathbf{D}_a$ *and axiom* 4b P *implies* **DP**.

Proof: Let σ be the translation which sends Q into Q' (if we wish to prove $\mathbf{D}_a$) or the dilatation with P as fixed point which sends Q into Q' (if we wish to prove **DP**). The lines l_1, l_2, l_3 will be among the traces of σ. Since

$$Q + R \parallel \sigma Q + \sigma R = Q' + \sigma R,$$

we obtain $\sigma R = R'$ and similarly $\sigma S = S'$.
Since

$$R + S \parallel \sigma R + \sigma S = R' + S'$$

our theorem is true.

THEOREM 2.17. $\mathbf{D}_a$ *implies axiom* 4a *and* **DP** *axiom* 4b P.

Proof:

1) Suppose each line contains only two points. Then there are two lines in a given pencil of parallel lines, hence only four points A, B, C, D in our geometry. There are six lines $A + B$, $A + C$, $\cdots$ each one containing only the two points by which it is expressed. Therefore this geometry must have a unique structure. On the other hand there is a field k with only two elements 0, 1 and the rule $1 + 1 = 0$. The geometry based on this field has only two points on a line and is therefore the four point geometry. This proves that all axioms hold in this four point geometry, especially axioms 4a and 4b. We may, therefore, assume that each line contains at least three points and it follows that to two lines one still can find a point not on these lines.

2) Assume $\mathbf{D}_a$. With any pair Q, Q' of distinct points we are going to associate a map $\tau^{Q,Q'}$ which will be defined only for points R not on the line $Q + Q'$.

Let $l \parallel Q + Q'$ and l contain R; then $l \neq Q + Q'$.

Let $m \parallel Q + R$ and m contain Q'; then $m \neq Q + R$.

Since $Q + Q' \nparallel Q + R$ we have $l \nparallel m$. Let R' be the point of intersection of m and l (Figure 2).

R' lies on l, hence not on $Q + Q'$, it lies on m, hence not on $Q + R$. This shows $R' \neq Q'$ and $R' \neq R$.

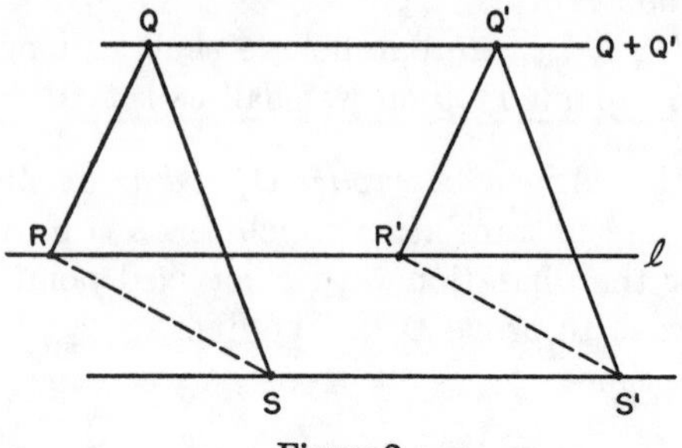

Figure 2

The construction may be described in simpler terms: $R + R' \parallel Q + Q'$ and $Q + R \parallel Q' + R'$. These statements describe the image R' of R and $\tau^{Q,Q'}$.

With this pair R, R' we may now construct the map $\tau^{R,R'}$. It will obviously send Q into Q'. Let now S be a point not on $R + R'$ or $Q + Q'$ and S' the image of S under $\tau^{Q,Q'}$. We have: $R + R' \parallel Q + Q' \parallel S + S'$ and these three lines are distinct. Furthermore $Q + R \parallel Q' + R'$ and $Q + S \parallel Q' + S'$. Since we assume $\mathbf{D}_a$ we can imply $R + S \parallel R' + S'$. But the statements $R + R' \parallel S + S'$ and $R + S \parallel R' + S'$ mean now that S' is also the image of S under $\tau^{R,R'}$. The two maps $\tau^{Q,Q'}$ and $\tau^{R,R'}$ agree wherever both are defined. Since we assume that a point S outside $Q + Q'$ and $R + R'$ does exist we can also form $\tau^{S,S'}$ and know of the three maps $\tau^{Q,Q'}$, $\tau^{R,R'}$, $\tau^{S,S'}$ that they agree wherever two or all three of the maps are defined. The desired map τ is now the combination of the three maps: for any point T the image shall be the image under one of the three maps (for whichever it is defined). This map τ has the property that $\tau(Q) = Q'$ as we have remarked already when we introduced $\tau^{R,R'}$. If we can show that τ is a dilatation, then it is clear that τ will be a translation since all traces have turned out to be parallel and the proof will be complete (the case $Q = Q'$ is trivial—identity is the map).

Let U, V be two distinct points. One of the lines $Q + Q'$, $R + R'$, $S + S'$ will not contain U and V and we may assume that $Q + Q'$

is this line so that we may use $\tau^{Q,Q'}$ for computing images. If $U + V \parallel Q + Q'$, then $U + V$ contains also U' and V'. We may, therefore, assume that $U + V \nparallel Q + Q'$. Then U and V are in the same position as our previous R and S for which we have proved that $R + S \parallel R' + S'$. This finishes the proof.

3) If we assume **DP**, we can prove 4b P by a method completely analoguous to the previous case using the lines through P instead of lines parallel to $Q + Q'$. We can leave the details to the reader.

Because of Theorems 2.16 and 2.17 our geometry is frequently called Desarguian geometry, our plane a Desarguian plane. Desargues' theorem is the other geometric interpretation of the axioms 4a and 4b which we have promised. A third geometric interpretation shall be discussed later.

7. Pappus' theorem and the commutative law

One of the simple but fascinating results in foundations of geometry is the fact that one can find a simple geometric configuration which is equivalent with the commutative law for multiplication of our field k.

Select an arbitrary point P. An element $\alpha \neq 0$ of k can be obtained (Theorem 2.10) by one and only one dilatation σ_α with P as fixed point in the form

$$\tau^\alpha = \sigma_\alpha \tau \sigma_\alpha^{-1}.$$

If $\tau^\beta = \sigma_\beta \tau \sigma_\beta^{-1}$, then $\tau^{\alpha\beta} = (\tau^\beta)^\alpha = \sigma_\alpha \sigma_\beta \tau \sigma_\beta^{-1} \sigma_\alpha^{-1} = \sigma_\alpha \sigma_\beta \tau (\sigma_\alpha \sigma_\beta)^{-1}$ on one hand and $\tau^{\alpha\beta} = \sigma_{\alpha\beta} \tau \sigma_{\alpha\beta}^{-1}$ on the other. Since σ_α is uniquely determined by α we have $\sigma_{\alpha\beta} = \sigma_\alpha \sigma_\beta$, and this means that the

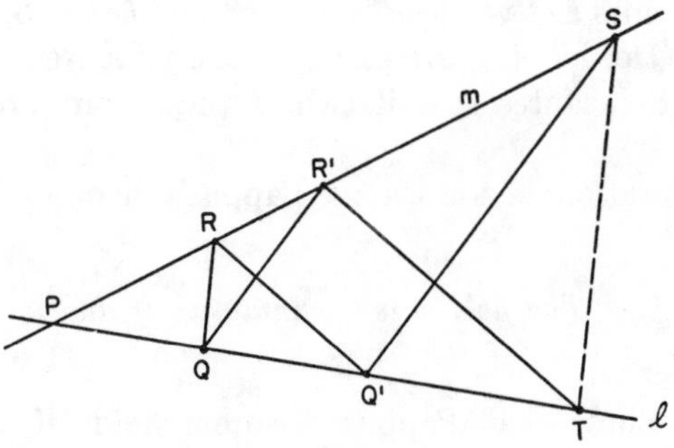

Figure 3

multiplicative group of non-zero elements of k is isomorphic to the group of dilatations which have P as fixed point. k will be commutative if and only if this group of dilatations is commutative.

Select now two distinct lines l and m through P and let Q be any point $\neq P$ on l. If σ_1 is a dilatation which has P as fixed point, then l is a σ_1-trace, $\sigma_1 Q = Q'$ will be $\neq P$ and may be any point on l by axiom 4b P. We describe, therefore, σ_1 completely by $\sigma_1 Q = Q'$. Similarly let us select points R, R' on m which are distinct from P and let us describe another dilatation σ_2 with P as fixed point by $\sigma_2 R = R'$ (Figure 3).

We shall first construct the two points $S = \sigma_1\sigma_2 R$ on m and $T = \sigma_2\sigma_1 Q$ on l. They are given completely by the descriptions:

$$Q + R' \parallel \sigma_1 Q + \sigma_1 R' = Q' + \sigma_1\sigma_2 R = Q' + S,$$

$$R + Q' \parallel \sigma_2 R + \sigma_2 Q' = R' + \sigma_2\sigma_1 Q = R' + T.$$

To have $\sigma_1\sigma_2 = \sigma_2\sigma_1$ it is necessary and sufficient to have $\sigma_1\sigma_2 Q = \sigma_2\sigma_1 Q$ or $\sigma_1\sigma_2 Q = T$. Since $\sigma_1\sigma_2 Q$ lies on l it is determined by

$$Q + R \parallel \sigma_1\sigma_2 Q + \sigma_1\sigma_2 R = \sigma_1\sigma_2 Q + S$$

and we see that $Q + R \parallel T + S$ is the condition for commutativity.

We can now forget σ_1 and σ_2 and look at the configuration. T on l and S on m are determined by

$$Q + R' \parallel Q' + S \quad \text{and} \quad Q' + R \parallel R' + T$$

and now we must have $Q + R \parallel T + S$. Six lines $Q + R'$, $R' + T$, $T + S$, $S + Q'$, $Q' + R$, $R + Q$ occur in this configuration which may be viewed as a hexagon "inscribed" in the pair l, m of lines with our six points as "vertices". $Q + R'$ and $Q' + S$, $Q' + R$ and $R' + T$, $R + Q$ and $T + S$ are pairs of opposite sides of the hexagon. The configuration states that if two of these pairs are parallel, so is the third.

This configuration is known as Pappus' theorem and we now may state

THEOREM 2.18. *The field k is commutative if and only if Pappus' theorem holds.*

EXERCISE. Show that Pappus' theorem holds if the two lines l and m are parallel and that for this case the commutativity of k

is *not* needed. Do not use a computation with coordinates. Make use of translations.

If a geometry contains only a finite number of points, then the group T is clearly finite and so will be the field k. If the reader has familiarized himself with Theorem 1.14 of Chapter I, §8 he will see the following geometric application.

THEOREM 2.19. *In a Desarguian plane with only a finite number of points the theorem of Pappus holds.*

No purely geometric proof is known for Theorem 2.19.

8. *Ordered geometry*

In some geometries, for instance in the case of the euclidean plane, the points of each line appear ordered. This ordering can, however, not be distinguished from the reversed ordering. An invariant way to describe such a situation is by means of a ternary relation: the point P lies "between" Q and R. Hilbert has axiomatized this ternary relation and the reader may consult his book "Grundlagen der Geometrie". We shall just follow the more naive way, considering a linear ordering and the reversed ordering as equivalent. We presuppose a knowledge of Chapter I, §9, at least Theorem 1.16.

The ordering must, however, be related to the geometry of the plane. The following definition describes this relation.

DEFINITION 2.8. A plane is said to be ordered if

1) the set of points on each line is linearly ordered;

2) parallel projection of the points of one line onto the points of another line either preserves or reverses the ordering.

The main result will be the following theorem (see Chapter I, §9, Definition 1.15).

THEOREM 2.20. *An ordering of a plane geometry induces canonically a weak ordering of the field k and a weak ordering of k induces canonically an ordering of the geometry.*

Proof:

1) Assume the geometry is ordered. We order the field k by the following procedure: Select a point P and a translation $\tau \neq 1$; then, as α ranges over k, the point $\tau^{\alpha}(P)$ will range over all points of the τ-trace l of P. We have, therefore, a one-to-one correspondence

between the elements α of k and the points $\tau^\alpha(P)$ of l. The ordering of k shall be the one induced by the ordering of l.

We must first of all show that this ordering does not depend on the selected P and τ.

Let us replace τ by a translation $\tau_1 \neq 1$ whose direction is different from that of τ and let l_1 be the τ_1-trace of P. We use again Theorem 2.10 stating that we can find for each non-zero element $\alpha \varepsilon k$ a unique dilatation σ_α which has P as fixed point and which satisfies $\tau_0^\alpha = \sigma_\alpha \tau_0 \sigma_\alpha^{-1}$ for all translations τ_0. Hence $\tau_0^\alpha(P) = \sigma_\alpha \tau_0 \sigma_\alpha^{-1}(P) = \sigma_\alpha \tau_0(P)$. We have, therefore,

$$\tau^\alpha(P) = \sigma_\alpha \tau(P) \quad \text{and} \quad \tau_1^\alpha(P) = \sigma_\alpha \tau_1(P).$$

$\tau(P)$ and $\tau^\alpha(P)$ are different from P and lie on l, $\tau_1(P)$ and $\tau_1^\alpha(P)$ are also different from P and lie on l_1. We have

$$\tau(P) + \tau_1(P) \parallel \sigma_\alpha \tau(P) + \sigma_\alpha \tau_1(P) = \tau^\alpha(P) + \tau_1^\alpha(P).$$

This means that the lines $\tau^\alpha(P) + \tau_1^\alpha(P)$ are all parallel—that $\tau_1^\alpha(P)$ is obtained from $\tau^\alpha(P)$ by parallel projection. This is, of course, also true if $\alpha = 0$ since this gives the point P which is the intersection of l and l_1. By our assumption the ordering of k induced by P and τ is either the same or the reverse of the one induced by P and τ_1.

If τ and τ_1 have the same direction, replace P, τ first by some P, τ_2 where τ_2 has a different direction and then P, τ_2 by P, τ_1. This shows the independence from τ.

Let us now replace P by a point Q and assume first that Q does not lie on l. Call l_1 the τ-trace of Q; the lines l and l_1 are parallel but distinct. We have

$$P + Q \parallel \tau^\alpha(P) + \tau^\alpha(Q)$$

which shows again that the points $\tau^\alpha(Q)$ are obtained from the points $\tau^\alpha(P)$ by parallel projection. The induced ordering is the same.

If Q lies on l, replace P first by a point Q_1 not on l and then Q_1 by Q. Our ordering of k does not depend on P and τ in the weak sense that it may possibly be reversed.

Now we must show that k is weakly ordered: Let δ be a fixed element of k and order k first by P and τ, then by $\tau^\delta(P)$ and τ. The two orderings will either be the same or reversed. In the first case the α will be ordered as the points $\tau^\alpha(P)$, in the second case as the points $\tau^\alpha(\tau^\delta(P)) = \tau^{\alpha+\delta}(P)$. This shows that the map $x \to x + \delta$ $(x \in k)$ either preserves or reverses the ordering. Order also, if $\delta \neq 0$, k first by P and τ and then by P and τ^δ. In the first case, the α are ordered as the points $\tau^\alpha(P)$, in the second case as the points $(\tau^\delta)^\alpha(P) = \tau^{\alpha\delta}(P)$. This shows that the map $x \to x\delta$ either preserves or reverses the ordering. k is indeed weakly ordered.

2) Assume now that k is weakly ordered. Let l be a given line. Select a point P on l and a translation $\tau \neq 1$ which has l as trace. $\tau^\alpha(P)$ will range over the points of l if α ranges over k. Order the points on l as the corresponding α are ordered. We have to show again that the ordering does not depend on the choices of P and τ.

Another point Q of l has the form $Q = \tau^\delta(P)$ for some fixed $\delta \in k$. Then $\tau^\alpha(Q) = \tau^{\alpha+\delta}(P)$. The ordering of the $\alpha + \delta$ is the same or the reverse of the ordering of the α and this shows the independence on the choice of Q. Let τ_1 be another translation $\neq 1$ with the same trace l. Then $\tau_1 = \tau^\delta$ with some $\delta \neq 0$. We find $\tau_1^\alpha(P) = (\tau^\delta)^\alpha(P) = \tau^{\alpha\delta}(P)$. The α and the $\alpha\delta$ are ordered either in the same or in the reversed way. This shows the independence of our prescription.

Finally we must show that parallel projection either preserves or reverses the ordering of the points on lines. Let l and m be distinct lines and suppose we are given a parallel projection mapping the points of l onto the points of m. We have to distinguish two cases:

a) l and m are parallel. Then we may select the same τ for both lines. Select P on l but select Q on m to be the image of P under the given parallel projection. The points of l are ordered according to $\tau^\alpha(P)$, those of m according to $\tau^\alpha(Q)$. But

$$P + Q \parallel \tau^\alpha(P) + \tau^\alpha(Q)$$

which means that $\tau^\alpha(Q)$ is the image of $\tau^\alpha(P)$ under our parallel projection.

b) l and m intersect in P. Let τ be any translation $\neq 1$ with trace l and Q the image of P under τ. Q will be a point $\neq P$ on the line l. Under our parallel projection it will have an image Q' on m which is $\neq P$. There is a translation τ_1 which carries P into Q'. Its trace through P will be m. Order the points of l by $\tau^\alpha(P)$, the points of m by $\tau_1^\alpha(P)$. As in 1) we have

$$\tau^\alpha(P) = \sigma_\alpha \tau(P) = \sigma_\alpha(Q) \quad \text{and} \quad \tau_1^\alpha(P) = \sigma_\alpha \tau_1(P) = \sigma_\alpha(Q')$$

if $\alpha \neq 0$. Then

$$Q + Q' \parallel \sigma_\alpha(Q) + \sigma_\alpha(Q') = \tau^\alpha(P) + \tau_1^\alpha(P)$$

which shows that $\tau_1^\alpha(P)$ is the image of $\tau^\alpha(P)$ by our parallel projection; this is trivially true if $\alpha = 0$. We see that the ordering of m corresponds to the ordering of l under our parallel projection.

3) Notice that the way to go from a weakly ordered field to an ordered geometry and the reverse procedure are consistent with each other.

The geometry with only four points is a freak case. Each line contains only two points, it is ordered trivially since only one ordering and its reverse are possible. This geometry corresponds to the field with two elements which is a freak case of a weakly ordered field.

Chapter I, §9, Theorem 1.16 implies now

THEOREM 2.21. *Aside from the plane with only* 4 *points and the field with only two elements the following is true: Any ordering of a Desarguian plane induces canonically an ordering of its field k and conversely each ordering of k induces an ordering of the plane.*

We shall leave aside the freak case.

DEFINITION 2.9. An ordering of a Desarguian plane is called archimedean if it has the following property:

If $\tau_1 \neq 1$, $\tau_2 \neq 1$ are translations with the same direction and if a point P does not lie between $\tau_1(P)$ and $\tau_2(P)$, then there exists an integer $n > 0$ such that $\tau_2(P)$ lies between P and $\tau_1^n(P)$.

Let us translate what this means for the field. We can write $\tau_2 = \tau_1^\alpha$. The three points $\tau_1^0(P) = P$, $\tau_1(P)$ and $\tau_2(P) = \tau_1^\alpha(P)$ are ordered as are $0, 1, \alpha$. The assumption is that 0 does not lie between 1 and α so that $\alpha > 0$. We wish to find an integer n such that $\tau_1^\alpha(P)$ lies between $\tau_1^0(P)$ and $\tau_1^n(P)$. This means $0 < \alpha < n$. By definition 1.16 of Chapter I, §9 this means that k is archimedean. Using Theorem 1.18 of Chapter I, §9 we see

THEOREM 2.22. *The necessary and sufficient condition for an ordered geometry to come from a field k which is isomorphic to a subfield of the field of real numbers (in its natural ordering) is the archimedean postulate. Notice, therefore, that in an archimedean plane the theorem of Pappus holds—the field is necessarily commutative.*

Chapter I, §9, Theorem 1.17 shows that one can not in general prove Pappus' theorem in an ordered geometry.

THEOREM 2.23. *There do exist ordered Desarguian planes in which the theorem of Pappus does not hold. These geometries are, of course, by necessity non-archimedean.*

EXERCISE. A consequence of the ordering axioms is the fact that a map $x \to \alpha x$ ($\alpha \neq 0$) either preserves or reverses the ordering. Give a geometric interpretation.

9. Harmonic points

Let Π be a Desarguian plane. We shall describe a certain configuration (Figure 4):

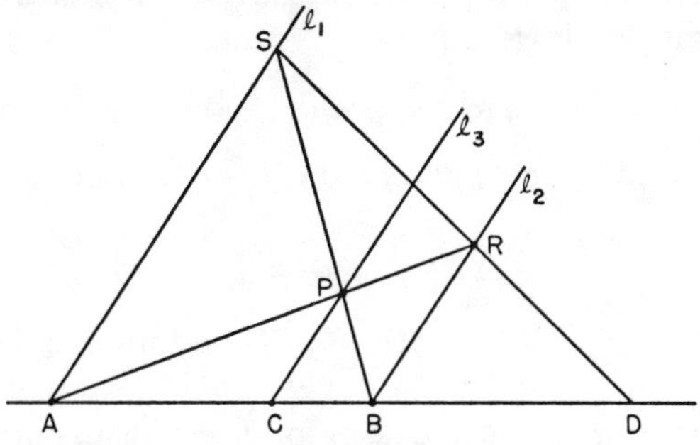

Figure 4

To three given distinct points A, B, C on a line l select three parallel lines l_1, l_2, l_3 which go respectively through A, B, C and are not parallel to l. Select on l_3 a point $P \neq C$. The arbitrary elements are, therefore, the pencil π of parallel lines containing l_1, l_2, l_3 but not l and the point $P \neq C$ on l_3.

The line $P + A$ is not $\parallel l_2$; otherwise it would be the line l_3 and A does not lie on l_3. The lines $P + A$ and l_2 intersect in a point R which is not on l; otherwise it would have to be the point B, and B does not lie on $P + A$ since P is not on l. Similarly $P + B$ will meet the line l_1 in a point S which is not on l. The lines l and $R + S$ are certainly distinct. If they are not parallel, they will meet in a point D; if they are parallel, we shall indicate this fact by putting $D = \infty$.

The contention is that D will not depend on the selected pencil π and on the point P. We shall prove this by computing the coordinates of D since we shall need this computation for a later purpose.

The origin of our coordinate system shall be a given point O on l and τ_1 a given translation $\neq 1$ with l as trace. This alone determines already the coordinates of any point X on $l : X = (\xi, 0)$ where ξ is determined by $\tau_1^{\xi}(O) = X$. We have, therefore, the freedom to select τ_2 as we please (without changing the coordinates of l) provided $\tau_2 \neq 1$, has not the trace l. We select τ_2 as the translation which carries the point C into the point P.

Let $(a, 0)$, $(b, 0)$, $(c, 0)$ be the coordinates of A, B, C. The map τ_1^c carries O into C and $\tau_2\tau_1^c$ carries O into P. The point P has, therefore, the coordinates $(c, 1)$. The elements a, b, c are distinct. l_1, l_2, l_3 are traces of τ_2. Any point on l_1 will have coordinates (a, η) since $\tau_2^{\eta}\tau_1^{a}(O) = \tau_2^{\eta}(A)$; a point on l_2 will have coordinates (b, η'). The equation of the line $P + A$ is

$$y = (x - a)(c - a)^{-1}$$

since it is satisfied by $(c, 1)$ and $(a, 0)$. $R = (b, \eta)$ lies on this line and we find

$$\eta = (b - a)(c - a)^{-1}.$$

Thus we have $R = (b, (b - a)(c - a)^{-1})$ and we find, by symmetry, $S = (a, (a - b)(c - b)^{-1})$.

Let $y = xm + n$ be the equation for $R + S$ (it is not of the form $x = e$ since $R + S$ is not $\parallel l_1$). The points R and S lie on it and we infer:

$$(2.6) \qquad \begin{aligned} (b - a)(c - a)^{-1} &= bm + n, \\ (a - b)(c - b)^{-1} &= am + n. \end{aligned}$$

If we subtract, the factor $b - a$ appears on both sides and we obtain

$$(2.7) \qquad m = (c - a)^{-1} + (c - b)^{-1}.$$

We stop for a moment and ask whether m can be 0. This would imply

$$(c - a) = -(c - b)$$

or $2c = a + b$. If the characteristic of k is $\neq 2$ this happens if and only if $c = \frac{1}{2}(a + b)$, if C is the "midpoint" of A and B. If the characteristic is two, then $0 = a + b$ or $a = b$ which is against our assumption; therefore $m = 0$ if and only if the characteristic of k is $\neq 2$ and if $c = \frac{1}{2}(a + b)$. For $m = 0$ the equation of $R + S$ is $y = n$;

it means that l and $R + S$ are parallel, hence $D = \infty$. Thus the case where $D = \infty$ does not depend on π and P.

Suppose that $m \neq 0$, $D = (d, 0)$. Since D lies on $y = xm + n$ we obtain

$$d = -nm^{-1}.$$

If we multiply (2.6) by m^{-1} we get

$$(2.8) \qquad (a - b)(c - b)^{-1}m^{-1} = a - d.$$

Let us first compute the inverse of the left side:

$$\begin{aligned} m(c - b)(a - b)^{-1} &= ((c - a)^{-1} + (c - b)^{-1})(c - b)(a - b)^{-1} \\ &= (c - a)^{-1}(c - b)(a - b)^{-1} + (a - b)^{-1} \\ &= (c - a)^{-1}((c - a) + (a - b))(a - b)^{-1} + (a - b)^{-1} \\ &= (a - b)^{-1} + (c - a)^{-1} + (a - b)^{-1} \\ &= 2(a - b)^{-1} + (c - a)^{-1}. \end{aligned}$$

Hence, from (2.8):

$$a - d = (2(a - b)^{-1} + (c - a)^{-1})^{-1}$$

and finally

$$(2.9) \qquad d = a - (2(a - b)^{-1} + (c - a)^{-1})^{-1}.$$

This result proves, of course, more than we have contended; notice that the expression for d uses only addition, subtraction and inverting.

Suppose that the points of our geometry are the ordered pairs (ξ, η) from a given field k. We have written the equation of a straight line in the form $xa + yb + c = 0$ with the coefficients on the right. The reason for this can be retraced; it originates from the fact that all our mappings were written on the left of the object we wanted to map. Suppose that we had done it the other way around. Then we would also have obtained a field $\bar{k}$ connected with the geometry. This field would now not be the the same as the field k, it would be an anti-isomorphic replica of k, the multiplication reversed. If, on the other hand, we start with a given field and call the ordered pairs points, we may introduce two geometries which—in general—

will not be the same. In one of them lines are given by equations $xa + yb + c = 0$ (let us call them "right lines"), in the other by equations $ax + by + c = 0$. Only a few lines will coincide for instance the x-axis: $y = 0$, most of the others will not (unless the field is commutative).

If we are given the three points $(a, 0)$, $(b, 0)$, $(c, 0)$ on the x-axis we can construct D in the "left" geometry or in the "right" geometry over our fixed field k. It is pretty obvious that we would end up with the same value (2.9) for d since multiplication of two elements is not involved in the formula.

We shall call A, B, C, D harmonic points and D the fourth harmonic point to A, B, C. Then D is the same in the left and in the right geometry. This is the first reason why we have written d in the apparently clumsy form (2.9).

Is D distinct from A, B, C? $D = A$ would mean that $R + S$ contains A. Since $S \neq A$, $R + S = l_1$ which is false. By symmetry $D \neq B$. How about $D = C$?

If the characteristic of k is 2, then (2.9) shows

$$d = a - ((c - a)^{-1})^{-1} = a - (c - a) = c$$

and we have indeed $D = C$.

Let k have a characteristic $\neq 2$. If we had $d = c$, then we would get from (2.9)

$$\begin{aligned} c - a &= -(2(a - b)^{-1} + (c - a)^{-1})^{-1}, \\ (c - a)^{-1} &= -2(a - b)^{-1} - (c - a)^{-1}, \\ 2(c - a)^{-1} &= -2(a - b)^{-1}, \\ c - a &= -a + b, \qquad c = b, \end{aligned}$$

which is false.

THEOREM 2.24. *If the characteristic of k is 2, then the fourth harmonic point D coincides with C and this configuration characterizes the fact that k has characteristic 2. If k does not have characteristic 2 (for instance in an ordered geometry with more than four points) the four harmonic points A, B, C, D will be distinct.*

Assume now that the characteristic of k is $\neq 2$.

The construction itself exhibits a symmetry in A and B. D does not change if A and B are interchanged. But there are more symmetries—(2.9) can be changed into an equivalent form by transposing the term a and taking an inverse:

$$(d - a)^{-1} = -2(a - b)^{-1} - (c - a)^{-1}$$

which is equivalent with

$$(2.10) \qquad (d - a)^{-1} + (a - b)^{-1} = (a - c)^{-1} + (b - a)^{-1}.$$

Interchanging C, D the condition remains the same.

Consider now

$$x^{-1} + y^{-1} = x^{-1}(x + y)y^{-1} = y^{-1}(x + y)x^{-1}.$$

Using this identity on both sides of (2.10) in two different ways and taking care that the factors $(b - a)^{-1}$ and $(a - b)^{-1} = -(b - a)^{-1}$ are factored out on both sides of the equation either to the left or to the right, we obtain after a cancellation two more equivalent forms:

$$(2.11) \qquad (d - b)(d - a)^{-1} = -(b - c)(a - c)^{-1},$$

$$(2.12) \qquad (d - a)^{-1}(d - b) = -(a - c)^{-1}(b - c).$$

If we interchange A and C and at the same time B and D in (2.11) we obtain

$$(b - d)(b - c)^{-1} = -(d - a)(c - a)^{-1};$$

just shifting terms it becomes

$$(d - a)^{-1}(b - d) = -(c - a)^{-1}(b - c)$$

which differs only in sign from (2.12) and is, therefore, correct.

We see that one should talk of two harmonic pairs A, B and C, D since we have established all the symmetries. We have, of course, assumed $D \neq \infty$, a restriction, which will be easily removed in the next paragraphs.

If our geometry is ordered, the two harmonic pairs A, B and C, D separate each other. Assume this were not the case. Placing the coordinate axis appropriately and making use of the symmetries we could assume that $a < b < c < d$. But then the left side of (2.11) would be positive and the right side would be negative which is a contradiction.

Assume still that the characteristic of k is $\neq 2$. The notion of two harmonic pairs has introduced some structure in as poor a geometry as that of a one dimensional line of a plane. It is now very natural to ask how "rigid" this structure is, in other words, what symmetries it has.

Let σ be a one-to-one map of the line l onto itself which carries harmonic points (finite points) into harmonic points. First of all σ^{-1} will have the same property. To show this, let A', B', C', D' be harmonic points and A, B, C, D their images under σ^{-1}. If the fourth harmonic point D_1 of A, B, C is finite, then A', B', C', $\sigma(D_1)$ are harmonic and, therefore, $\sigma D_1 = D' = \sigma D$, $D_1 = D$. We are in doubt if $c = (a + b)/2$; but if it were, then $d \neq (a + b)/2$ and we could conclude that the fourth harmonic point of A, B, D must be C. It follows now that three points whose fourth harmonic is at ∞ must have images with the same property, or else we would get a contradiction using σ^{-1}. This shows that

$$\sigma\left(\frac{a+b}{2}\right) = \frac{\sigma(a) + \sigma(b)}{2} \tag{2.13}$$

whenever $a \neq b$ (considering the triplet a, b, $(a + b)/2$). If $a = b$ the equality is trivial.

Make at first the special assumption that $\sigma(0) = 0$ and $\sigma(1) = 1$. For $b = 0$, (2.13) becomes

$$\sigma\left(\frac{a}{2}\right) = \frac{\sigma(a)}{2}.$$

This can be used on (2.13):

$$\sigma(a + b) = \sigma(a) + \sigma(b) \tag{2.14}$$

which means that σ is an isomorphism under addition. Since $\sigma(1) = 1$ we get $\sigma(-1) = -1$.

Let $a = -1$, $b = +1$ and $c \neq 0, 1, -1$. From (2.9) we obtain for the fourth harmonic point

$$d = -1 - (-1 + (c + 1)^{-1})^{-1}.$$

Since $-1 + (c + 1)^{-1} = (-(c + 1) + 1)(c + 1)^{-1} = -c(c + 1)^{-1}$ we have

$$d = -1 + (c + 1)c^{-1} = -1 + 1 + c^{-1} = c^{-1}.$$

In other words -1, 1, c, c^{-1} are harmonic. Their images -1, 1, $\sigma(c)$, $\sigma(c^{-1})$ are again harmonic showing

$$\sigma(c^{-1}) = (\sigma(c))^{-1}, \tag{2.15}$$

an equation which is trivially true for $c = -1$ so that only $c \neq 0$ is needed. Conversely, if a map satisfies (2.14) and (2.15), then one glance at formula (2.9) tells us that if d is the fourth harmonic point of a, b, c, then $\sigma(d)$ will be the fourth harmonic point of $\sigma(a)$, $\sigma(b)$, $\sigma(c)$, since only sums, differences and inverses are involved in (2.9). Conditions (2.14) (2.15) together with $\sigma(1) = 1$ characterize such maps.

We turn now to Chapter I, §8, Theorem 1.15 and see that σ is either an automorphism or an antiautomorphism of k.

To settle the general case where $\sigma(0) = 0$ or $\sigma(1) = 1$ may be violated, notice that $\sigma(x) - \sigma(0)$ differs from $\sigma(x)$ by a translation and $(\sigma(1) - \sigma(0))^{-1}(\sigma(x) - \sigma(0)) = \tau(x)$ differs from $\sigma(x) - \sigma(0)$ by a dilatation. Dilatations preserve the configuration of harmonic points. But $\tau(0) = 0$, $\tau(1) = 1$ so that $\tau(x) = x^\tau$ is either an automorphism or an antiautomorphism of k. Computing $\sigma(x)$ we find $\sigma(x) = (\sigma(1) - \sigma(0))x^\tau + \sigma(0)$; it is, therefore, of the form $\sigma(x) = ax^\tau + b$ with $a \neq 0$.

THEOREM 2.25. *The one-to-one onto maps σ of a line onto itself which preserve harmonic (finite) pairs are of the form*

$$\sigma(x) = ax^\tau + b$$

where $a \neq 0$ and where τ is either an automorphism or an antiautomorphism of k.

REMARK. If identity is the only automorphism of k (k must be commutative in such a case, since otherwise there would be inner automorphisms), for instance, if k is the field of real numbers, then $\sigma(x) = ax + b$. This result was first obtained by von Staudt.

10. *The fundamental theorem of projective geometry*

If V is a left vector space over a field k and V' a left vector space over a field k' and if there exists an isomorphism μ which maps k onto k', then we can generalize the notion of a homomorphism in an obvious fashion.

DEFINITION 2.10. A map $\lambda : V \to V'$ is called semi-linear with respect to the isomorphism μ if

1) $\lambda(X + Y) = \lambda(X) + \lambda(Y)$, 2) $\lambda(aX) = a^\mu\lambda(X)$ for all $X, Y \in V$ and all $a \in k$.

The description of all semi-linear maps follows the same pattern as that of homomorphisms. If $A_1, A_2, \cdots, A_n$ is a basis of V, if $B_1, B_2, \cdots, B_n$ are arbitrary vectors in V' (thought of as images of the basis vectors A_i) and if $X = \sum_i x_i A_i$ is an arbitrary vector of V then

$$\lambda(X) = \sum_{i=1}^{n} x_i^\mu B_i$$

is a semi-linear map of V into V' and every semi-linear map is of this form. It is one-to-one if and only if the B_i are independent, and one-to-one and onto V' if and only if the B_i are a basis of V'. We can leave these rather trivial details to the reader.

With each left vector space V over k we form a new object, the corresponding projective space $\bar{V}$. Its elements are no longer the single vectors of V, they are the subspaces U of V. To each subspace U of V we assign a projective dimension: $\dim_p U = \dim U - 1$, just by 1 smaller than the ordinary dimension.

We take over the terminology "points" and "lines" for subspaces with projective dimension 0, respectively 1. Thus the lines of V become the "points" of $\bar{V}$ and the planes of V become the "lines" of $\bar{V}$. The whole space V of ordinary dimension n gets (as element of $\bar{V}$) the projective dimension $n - 1$. The 0-subspace of V should be thought of as the "empty element" of $\bar{V}$ and has projective dimension -1. An incidence relation is introduced in $\bar{V}$; it shall merely be the inclusion relation $U_1 \subset U_2$ between the subspaces of V. Intersection $U_1 \cap U_2$ and sum $U_1 + U_2$ can then be explained in terms of the incidence relation of $\bar{V}$: $U_1 \cap U_2$ is the "largest" element contained in U_1 and U_2, $U_1 + U_2$ is the "smallest" element which contains both U_1 and U_2.

Definition 2.11. A map $\sigma : \bar{V} \to \overline{V'}$ of the elements of a projective space $\bar{V}$ onto the elements of a projective space $\overline{V'}$ is called a collineation if 1) $\dim V = \dim V'$, 2) σ is one-to-one and onto, 3) $U_1 \subset U_2$ implies $\sigma U_1 \subset \sigma U_2$.

As an example of a collineation suppose that there exists a semi-linear map $\lambda : V \to V'$ which is one-to-one and onto. If we define $\sigma U = \lambda U$, then σ is obviously a collineation of $\bar{V} \to \overline{V'}$ and we say that σ is induced by λ. The main purpose of this paragraph is to prove that every collineation is induced by some semi-linear map if dim $V \geq 3$.

Let $\sigma : \bar{V} \to \overline{V'}$ be a collineation, U a subspace of V and dim $V = n$. We can find a sequence $0 = U_0 \subset U_1 \subset U_2 \subset \cdots \subset U_n = V$ of subspaces, where dim $U_i = j$, such that U is one of the terms. By assumption: $\sigma U_0 \subset \sigma U_1 \subset \cdots \subset \sigma U_n$. Since $U_i \neq U_{i+1}$ and since σ was one-to-one it follows that $\sigma U_i \neq \sigma U_{i+1}$. The dimensions of the σU_i increase, therefore, strictly with i. But dim $V' = n$ puts an upper bound n on dim σU_i. This implies dim $\sigma U_i = i$ and consequently dim $U =$ dim σU. A collineation preserves the dimension.

Let U_1 and U_2 be subspaces of V and suppose that $\sigma U_1 \subset \sigma U_2$. We can then find a supplementary space to σU_1 in σU_2; it must be the image of some W, in other words $\sigma U_2 = \sigma U_1 + \sigma W$ with $\sigma U_1 \cap \sigma W = 0$. Since $U_1 \cap W$ is a subspace of U_1 and of W we conclude that $\sigma(U_1 \cap W) \subset \sigma U_1 \cap \sigma W = 0$. This shows $U_1 \cap W = 0$. Therefore

$$\dim(U_1 + W) = \dim U_1 + \dim W = \dim \sigma U_1 + \dim \sigma W$$
$$= \dim(\sigma U_1 + \sigma W) = \dim(\sigma U_2),$$

hence

$$\dim(\sigma(U_1 + W)) = \dim \sigma U_2 .$$

On the other hand U_1 as well as W are subspaces of $U_1 + W$ and consequently $\sigma U_1 + \sigma W \subset \sigma(U_1 + W)$ or $\sigma U_2 \subset \sigma(U_1 + W)$. Since the dimensions are equal we get $\sigma U_2 = \sigma(U_1 + W)$ and since σ is one-to-one we have $U_2 = U_1 + W$; therefore finally $U_1 \subset U_2$. All this was needed to show that σ^{-1} is also a collineation. Since σ as well as σ^{-1} preserves the inclusion relation and since intersection and sum can be explained in terms of the inclusion relation we obtain also that

$$\sigma(U_1 \cap U_2) = \sigma U_1 \cap \sigma U_2 \quad \text{and} \quad \sigma(U_1 + U_2) = \sigma U_1 + \sigma U_2 .$$

Suppose that we know of σ only the effect on the "points" of $\bar{V}$ (they are the lines of V). If U is any subspace of V and $U = L_1 + L_2 + \cdots + L_r$ where each L_i is a line, then $\sigma U = \sigma L_1 + \sigma L_2 + \cdots + \sigma L_r$ and we see that σ is completely known if its effect on the lines of V is known. Its effect on lines will have the following property: Whenever $L_1 \subset L_2 + L_3$ then $\sigma L_1 \subset \sigma L_2 + \sigma L_3$. We can now formulate and prove the fundamental theorem of projective geometry:

THEOREM 2.26. (*Fundamental theorem of projective geometry*). *Let V and V' be left vector spaces of equal dimension $n \geq 3$ over fields k respectively k', $\bar{V}$ and $\overline{V'}$ the corresponding projective spaces. Let σ be a one-to-one (onto) correspondence of the "points" of $\bar{V}$ and the "points" of $\overline{V'}$ which has the following property: Whenever three distinct "points" L_1, L_2, L_3 (they are lines of V) are collinear: $L_1 \subset L_2 + L_3$, then their images are collinear: $\sigma L_1 \subset \sigma L_2 + \sigma L_3$. Such a map can of course be extended in at most one way to a collineation but we contend more. There exists an isomorphism μ of k onto k' and a semi-linear map λ of V onto V' (with respect to μ) such that the collineation which λ induces on $\bar{V}$ agrees with σ on the points of $\bar{V}$. If λ_1 is another semi-linear map with respect to an isomorphism μ_1 of k onto k' which also induces this collineation, then $\lambda_1(X) = \lambda(\alpha X)$ for some fixed $\alpha \neq 0$ of k and the isomorphism μ_1 is given by $x^{\mu_1} = (\alpha x \alpha^{-1})^{\mu}$. For any $\alpha \neq 0$ the map $\lambda(\alpha X)$ will be semi-linear and induce the same collineation as λ. The isomorphism μ is, therefore, determined by σ up to inner automorphisms of k.*

REMARK 1. The assumption that L_1, L_2, L_3 are distinct is irrelevant since $L_1 \subset L_2 + L_3$ will imply trivially that $\sigma L_1 \subset \sigma L_2 + \sigma L_3$ as soon as two of the "points" are equal.

REMARK 2. If $n = 2$, then $\dim_r V = 1$. There is only one line, and a completely random one-to-one correspondence of the points of $\bar{V}$ and the points of $\overline{V'}$ would be a collineation. The statement would be false for any field with at least 5 elements; only for F_2, F_3, F_4 would it still hold.

Proof:

1) We show by induction on r: if $L \subset L_1 + L_2 + \cdots + L_r$, then $\sigma L \subset \sigma L_1 + \sigma L_2 + \cdots + \sigma L_r$. This is trivial if $L = L_r$. If $L \neq L_r$, then L is spanned by a vector $A + B$ where $A \in L_1 + L_2 + \cdots + L_{r-1}$ ($A \neq 0$) and $B \in L_r$. Then $\sigma\langle A\rangle \subset \sigma L_1 + \sigma L_2 + \cdots + \sigma L_{r-1}$ by induction hypothesis and from $L \subset \langle A\rangle + L_r$ we get $\sigma L \subset \sigma\langle A\rangle + \sigma L_r$.

2) We shall use frequently the following reasoning. Let C and D

be independent vectors and $L \subset \langle C\rangle + \langle D\rangle$ but $L \neq \langle D\rangle$. Then L is spanned by a vector $aC + bD$, with $a \neq 0$, which may be replaced by $C + a^{-1}bD = C + dD$. The element d is then uniquely determined by L.

3) Let A_i be a basis of V, $L_i = \langle A_i\rangle$, $\sigma L_i = \langle A_i'\rangle$. We have $V = L_1 + L_2 + \cdots + L_n$; if L is any line of V then $L \subset V$ and consequently $\sigma L \subset \sigma L_1 + \sigma L_2 + \cdots + \sigma L_n$. Since our map is onto, any line of V' will have the form σL which shows $\sigma L_1 + \sigma L_2 + \cdots + \sigma L_n = V'$ so that the A_i' span V'. But $\dim V' = n$ implies that the A_i' are a basis of V'. The line $\langle A_1 + A_i\rangle$ will be $\subset \langle A_1\rangle + \langle A_i\rangle$ and distinct from $\langle A_i\rangle$ if $i \geq 2$; hence $\sigma\langle A_1 + A_i\rangle = \langle A_1' + b_iA_i'\rangle$; since $\langle A_1 + A_i\rangle \neq \langle A_1\rangle$ we will have $b_i \neq 0$. Let us replace A_i' by the equivalent vector b_iA_i'. Then $\sigma\langle A_i\rangle = \langle A_i'\rangle$ for $i \geq 1$ and $\sigma\langle A_1 + A_i\rangle = \langle A_1' + A_i'\rangle$ for $i \geq 2$.

4) Let $x \in k$; $\langle A_1 + xA_2\rangle \subset \langle A_1\rangle + \langle A_2\rangle$ and $\langle A_1 + xA_2\rangle \neq \langle A_2\rangle$. There is a uniquely determined $x' \in k'$ such that $\sigma\langle A_1 + xA_2'\rangle = \langle A_1' + x'A_2'\rangle$ and, since $\langle A_1 + xA_2\rangle \neq \langle A_1 + yA_2\rangle$ if $x \neq y$, we have $x' \neq y'$. The map $k \to k'$ given by $x \to x'$ is, therefore, one-to-one but only into, thus far. Clearly $0' = 0$ and $1' = 1$ by 3).

In a similar way we could find a map $x \to x''$ such that $\sigma\langle A_1 + xA_3\rangle = \langle A_1' + x''A_3'\rangle$. We contend now that $x' = x''$ for all $x \in k$. We can assume $x \neq 0$.

The line $\langle xA_2 - xA_3\rangle$ lies in $\langle A_2\rangle + \langle A_3\rangle$ on one hand and in $\langle A_1 + xA_2\rangle + \langle A_1 + xA_3\rangle$ on the other. Its image is, therefore, spanned by a vector of $\langle A_2'\rangle + \langle A_3'\rangle$ and also by a vector of $\langle A_1' + x'A_2'\rangle + \langle A_1' + x''A_3'\rangle$. The only possibility for the image is the line $\langle x'A_2' - x''A_3'\rangle$. But $\langle xA_2 - xA_3\rangle = \langle A_2 - A_3\rangle$ whose image is by the same reasoning $\langle 1'A_2' - 1''A_3'\rangle = \langle A_2' - A_3'\rangle$. This shows $\langle x'A_2' - x''A_3'\rangle = \langle A_2' - A_3'\rangle$ and $x'' = x'$ follows.

Instead of A_3 we could have taken any vector A_i with $i \geq 3$. We conclude that there is but one map $x \to x'$ such that, for $i \geq 2$,

$$\sigma\langle A_1 + xA_i\rangle = \langle A_1' + x'A_i'\rangle.$$

5) Suppose $\sigma\langle A_1 + x_2A_2 + \cdots + x_{r-1}A_{r-1}\rangle = \langle A_1' + x_2'A_2' + \cdots + x_{r-1}'A_{r-1}'\rangle$ is proved. The line $\langle A_1 + x_2A_2 + \cdots + x_rA_r\rangle$ lies in $\langle A_1 + x_2A_2 + \cdots + x_{r-1}A_{r-1}\rangle + \langle A_r\rangle$ and is distinct from $\langle A_r\rangle$. Its image is, therefore, spanned by a vector of the form $A_1' + x_2'A_2' + \cdots + x_{r-1}'A_{r-1}' + uA_r'$. Our line is also in $\langle A_1 + x_rA_r\rangle + \langle A_2\rangle + \cdots + \langle A_{r-1}\rangle$ which implies that the image is in $\langle A_1' + x_r'A_r'\rangle +$

$\langle A_2' \rangle + \cdots + \langle A_{r-1}' \rangle$. Since the image must use A_1' we get now $u = x_r'$. We know, therefore,

$$\sigma\langle A_1 + x_2A_2 + \cdots + x_nA_n \rangle = \langle A_1' + x_2'A_2' + \cdots + x_n'A_n' \rangle.$$

6) The image of the line $\langle x_2A_2 + \cdots + x_nA_n \rangle$ is in $\langle A_2' \rangle + \cdots + \langle A_n' \rangle$. It also lies in $\langle A_1 + x_2A_2 + \cdots + x_nA_n \rangle + \langle A_1 \rangle$, its image, therefore, in $\langle A_1' + x_2'A_2' + \cdots + x_n'A_n' \rangle + \langle A_1' \rangle$. Clearly $\sigma\langle x_2A_2 + \cdots + x_nA_n \rangle = \langle x_2'A_2' + \cdots + x_n'A_n' \rangle$.

7) A given line of V' which has the form $\langle A_1' + yA_2' \rangle$ is not the image of a line in $\langle A_2 \rangle + \cdots + \langle A_n \rangle$ as 6) has shown. Hence it must be the image of a line $\langle A_1 + x_2A_2 + \cdots + x_nA_n \rangle$, which implies $x_2' = y$ and shows that the map $k \to k'$ is onto.

8) $\sigma\langle A_1 + (x + y)A_2 + A_3 \rangle = \langle A_1' + (x + y)'A_2' + A_3' \rangle$.

But

$$\langle A_1 + (x + y)A_2 + A_3 \rangle \subset \langle A_1 + xA_2 \rangle + \langle yA_2 + A_3 \rangle$$

so that

$$\langle A_1' + (x + y)'A_2' + A_3' \rangle \subset \langle A_1' + x'A_2' \rangle + \langle y'A_2' + A_3' \rangle.$$

We deduce easily $(x + y)' = x' + y'$.

9) $\sigma\langle A_1 + xyA_2 + xA_3 \rangle = \langle A_1' + (xy)'A_2' + x'A_3' \rangle$

and

$$\langle A_1 + xyA_2 + xA_3 \rangle \subset \langle A_1 \rangle + \langle yA_2 + A_3 \rangle$$

hence

$$\langle A_1' + (xy)'A_2' + x'A_3' \rangle \subset \langle A_1' \rangle + \langle y'A_2' + A_3' \rangle$$

which implies $(xy)' = x'y'$. Our map $x \to x'$ is an isomorphism μ of k onto k'.

The line $\langle x_1A_1 + \cdots + x_nA_n \rangle$ has $\langle x_1^{\mu}A_1' + \cdots + x_n^{\mu}A_n' \rangle$ as image; this is clear if $x_1 = 0$ or $x_1 = 1$. If $x_1 \neq 0$ our line is also given by $\langle A_1 + x_1^{-1}x_2A_2 + \cdots + x_1^{-1}x_nA_n \rangle$ and its image $\langle A_1' + x_1^{-\mu}x_2^{\mu}A_2' + \cdots + x_1^{-\mu}x_n^{\mu}A_n' \rangle$ is the same line as $\langle x_n^{\mu}A_1' + \cdots + x_n^{\mu}A_n' \rangle$.

10) Let λ be the semi-linear map of V onto V' (with respect to the isomorphism μ) which sends A_i onto A_i'. For any line L of V we have shown $\sigma L = \lambda L$. This proves the first part of the theorem. The rest is very easy.

11) Suppose λ_1 is another semi-linear map of V onto V' which

has the same effect on the lines of V. Then $\lambda^{-1}\lambda_1$ is a map of V onto V which keeps every line of V fixed. It will map every non-zero vector X of V onto some non-zero multiple of X. Let X and Y be independent vectors of V. The three vectors X, Y, $X + Y$ will go respectively into αX, βY, $\gamma(X + Y)$. But $\lambda^{-1}\lambda_1(X + Y) = \lambda^{-1}\lambda_1(X) + \lambda^{-1}\lambda_1(Y) = \alpha X + \beta Y$. A comparison with $\gamma X + \gamma Y$ shows $\alpha = \beta$. If X, Y are dependent but $\neq 0$ and Z a third independent vector, then both X and Y take on the same factor, under $\lambda^{-1}\lambda_1$, as Z. It follows now that $\lambda^{-1}\lambda_1(X) = \alpha X$ with the same α for all X (the case that $X = 0$ is trivial) and consequently $\lambda_1(X) = \lambda(\alpha X)$.

12) Let $\alpha \neq 0$ and define the map λ_1 by $\lambda_1(X) = \lambda(\alpha X)$. For any $\beta \in k$ we have $\lambda_1(\beta X) = \lambda(\alpha\beta X) = \lambda(\alpha\beta\alpha^{-1}\cdot\alpha X) = (\alpha\beta\alpha^{-1})^{\mu}\cdot\lambda_1(X)$. The map λ_1 is, therefore, semi-linear and its isomorphism μ_1 is given by $x^{\mu_1} = (\alpha x\alpha^{-1})^{\mu}$. If $A \neq 0$, then $\lambda_1(A) = \lambda(\alpha A) = \alpha^{\mu}\lambda(A)$ which shows that $\lambda_1\langle A\rangle = \lambda\langle A\rangle$ so that λ_1 and λ induce the same collineation. The proof of our theorem is complete.

For any semi-linear map λ, let us denote its isomorphism by $\bar{\lambda}$. Suppose now $V \xrightarrow{\lambda_1} V' \xrightarrow{\lambda_2} V''$ is a succession of semi-linear maps. Then

$$\lambda_2\lambda_1(xX) = \lambda_2(x^{\bar{\lambda}_1}\lambda_1(X)) = (x^{\bar{\lambda}_1})^{\bar{\lambda}_2}\lambda_2\lambda_1(X) = x^{\bar{\lambda}_2\bar{\lambda}_1}\lambda_2\lambda_1(X).$$

We see that $\lambda_2\lambda_1$ is semi-linear again and that $\overline{\lambda_2\lambda_1} = \bar{\lambda}_2\bar{\lambda}_1$.

Very interesting results are obtained in the case of collineations of a projective space $\bar{V}$ onto itself. The semi-linear maps of V onto itself form a group S; the collineations of $\bar{V}$ form also a group which shall be denoted by PS. Each element λ of S induces a collineation of $\bar{V}$ onto itself. The map $S \xrightarrow{j} PS$ which associates with each $\lambda \in S$ the collineation $j(\lambda)$ induced by it ($j(\lambda)$ is merely the effect λ has on the projective space $\bar{V}$) is clearly a homomorphism of S onto PS; it is an onto map by the fundamental theorem of projective geometry. The kernel of j consists of those $\lambda \in S$ which keep every line of V fixed and we have seen that there is one such map λ_α for each $\alpha \neq 0$ of k, hence for each $\alpha \in k^*$, namely the map $\lambda_\alpha(X) = \alpha X$. We may then map k^* into S by $\alpha \xrightarrow{i} \lambda_\alpha$; since $\lambda_\alpha\lambda_\beta = \lambda_{\alpha\beta}$, this map i is an isomorphism of k^* *into* S and the image of i will be the kernel of j. It is customary to express such a situation by a diagram

$$1 \to k^* \xrightarrow{i} S \xrightarrow{j} PS \to 1$$

where two entirely trivial maps are added: every element of PS is mapped onto 1, and $1 \to k^*$ maps 1 onto the unit element of k^*.

In this succession the image of each map is the kernel of the next map; if this situation occurs one calls the sequence of maps an *exact* sequence. Thus the statement that our sequence is exact contains all the previous ones, i.e., that i is an isomorphism, that its image is the kernel of j and that j is onto.

Denote now by $A(k)$ the automorphisms of k. We can map $S \xrightarrow{\rho} A(k)$, where ρ denotes the map $\lambda \to \bar{\lambda}$ which associates with each semi-linear map λ the automorphism $\bar{\lambda}$. The map ρ is a homomorphism as we have seen and it is an onto map since every automorphism is possible in some semi-linear map.

The kernel GL (general linear group) of the map ρ consists of those elements of S for which the automorphism is identity. They are merely our old homomorphisms of V onto V. We have, therefore, another exact sequence

$$1 \to GL \xrightarrow{inj} S \xrightarrow{\rho} A(k) \to 1$$

where inj indicates the mere injection map of GL into S.

Each element of GL induces on $\bar{V}$ a special collineation called projective transformation of $\bar{V}$. These transformations form a subgroup of PS denoted by PGL (projective general linear group), and the map of GL onto PGL is just the restriction of j to GL. We shall (rather sloppily) call it j again. The kernel consists of those λ_α which are homomorphisms. They must satisfy $\lambda_\alpha(xX) = x\lambda_\alpha(X)$ or $\alpha x = x\alpha$ for all $x \in k$. This means that α comes from the center Z of k, hence from Z^*, since $\alpha \neq 0$. We get the exact sequence

$$1 \to Z^* \xrightarrow{i} GL \xrightarrow{j} PGL \to 1.$$

Denote by $I(k)$ the inner automorphisms of k. An element of PS determines an automorphism of k only up to an inner one so that we get only an element of the factor group $A(k)/I(k)$ associated with each element of PS. Let us denote by ρ' this onto map $PS \xrightarrow{\rho'} A(k)/I(k)$. Its kernel consists of the collineations which are induced by elements of S belonging to an inner automorphism. But since one can change the automorphism by any inner one we can assume that the element of S belongs to the identity automorphism, that it is in GL. This shows that PGL is the kernel of ρ' and another exact sequence may be written down.

The situation will become less confusing if the reader studies the following diagram which incorporates all the statements we have made and explains in full detail the consequences of the fundamental theorem of projective geometry in the case of collineations of the space $\bar{V}$ onto itself:

$$
\begin{array}{ccccccc}
 & & 1 & & 1 & & 1 & \\
 & & \downarrow & & \downarrow & & \downarrow & \\
1 \rightarrow & & Z^* & \xrightarrow{i} & GL & \xrightarrow{j} & PGL & \rightarrow 1 \\
 & & \downarrow{\scriptstyle inj} & & \downarrow{\scriptstyle inj} & & \downarrow{\scriptstyle inj} & \\
1 \rightarrow & & k^* & \xrightarrow{i} & S & \xrightarrow{j} & PS & \rightarrow 1 \\
 & & \downarrow{\scriptstyle \rho_0} & & \downarrow{\scriptstyle \rho} & & \downarrow{\scriptstyle \rho'} & \\
1 \rightarrow & & I(k) & \xrightarrow{inj} & A(k) & \xrightarrow{can} & A(k)/I(k) & \rightarrow 1 \\
 & & \downarrow & & \downarrow & & \downarrow & \\
 & & 1 & & 1 & & 1 &
\end{array}
$$

The symbol "inj" indicates a mere injection map, "can" the canonical map onto the factor group. The map ρ_0 associates with $\alpha \varepsilon k^*$ the inner automorphism $x \rightarrow \alpha x \alpha^{-1}$; its kernel is clearly Z^*.

Every row and every column is an exact sequence. Every square of the diagram is "commutative"; this means for instance for the square whose left upper corner is S that the map j followed by ρ' gives the same result as ρ followed by can, i.e., $\rho' j = \text{can } \rho$.

In later chapters we shall study subgroups Γ of GL. The effect which Γ has on the projective space $\bar{V}$, therefore the image of Γ under j, will be denoted by $P\Gamma$. The kernel of this onto map $\Gamma \rightarrow P\Gamma$ consists of those λ_α which belong to Γ. In all cases which we shall consider this kernel will be the center of Γ, and this will allow us to define $P\Gamma$ as the factor group of Γ modulo its center.

There is a geometric construction called projection (or perspectivity) which dominated the classical projective geometry and which will give us still a better understanding of the fundamental theorem of projective geometry.

Assume that both spaces V and V' are subspaces of a space Ω of dimension $N > \dim V = \dim V' = n \geq 2$; notice that we allow $n = 2$. Let T be a subspace of Ω which is supplementary to V as well as to V':

$$\Omega = V \oplus T = V' \oplus T, \qquad V \cap T = V' \cap T = 0.$$

Let W be a subspace of Ω which contains T and intersect W with V: $W \cap V = U$. Clearly $U + T \subset W$ but the reverse inclusion is also true. Any vector of $\Omega = V \oplus T$ has the form $A = B + C$ with $B \varepsilon V$ and $C \varepsilon T$. If $A \varepsilon W$, then $B = A - C$ lies in W and V, therefore in U; consequently $A \varepsilon U + T$ and we obtain $W = U + T$. If we start with any subspace U of V and set $W = U + T$, then $U \subset W \cap V$; if $A \varepsilon W \cap V \subset W$, we can write $A = B + C$ where $B \varepsilon U$ and $C \varepsilon T$. The vector $C = A - B$ lies in V as well as in T and, since $V \cap T = 0$, we get $C = 0$, $A = B \varepsilon U$.

For subspaces U of V and spaces W which contain T the two equations

$$U = W \cap V \quad \text{and} \quad W = U + T$$

are, therefore, equivalent.

We start now with any $U \subset V$, form $W = U + T$ and intersect W with V': $U' = W \cap V'$. This gives a one-to-one correspondence between the subspaces U of V and the subspaces U' of V' which is completely described by the one equation

$$U + T = U' + T.$$

Since $U_1 \subset U_2$ implies $U_1' \subset U_2'$ this one-to-one correspondence is a collineation of $\bar{V}$ onto $\overline{V'}$ which one calls projection of $\bar{V}$ onto $\overline{V'}$ with center $\bar{T}$.

We allow also $n = 2$ and can, therefore, not make use of the fundamental theorem of projective geometry. We proceed directly. Let $A_1, A_2, \cdots, A_n$ be a basis of V and denote by $\langle A_i' \rangle$ the image of the line $\langle A_i \rangle$. Then $\langle A_i \rangle + T = \langle A_i' \rangle + T$ and this leads to a formula $A_i = \alpha_i A_i' + B_i$ where $B_i \varepsilon T$; since $A_i \not\varepsilon T$ the element α_i is $\neq 0$. We may replace A_i' by the equivalent vector $\alpha_i A_i'$, thereby simplifying the formula to $A_i = A_i' + B_i$. This leads to $x_1A_1 + \cdots + x_nA_n = x_1A_1' + \cdots + x_nA_n' + C$ with $C \varepsilon T$. If we denote by λ the homomorphism (with respect to the identity automorphism of k) which sends A_i onto A_i', we can write $\lambda(X) - X \varepsilon T$ and obtain

$$\langle \lambda(X) \rangle + T = \langle X \rangle + T.$$

This formula shows that the linear map λ induces on $\bar{V}$ our projection. The map $\lambda : V \to V'$ is not an arbitrary homomorphism. Should $X \varepsilon V \cap V'$ then $\langle X \rangle + T = \langle X \rangle + T$ (in different interpretations) shows that $\langle X \rangle$ is kept fixed by the projection; $\lambda(X) - X$ is, therefore, not only in T but also in V and $V \cap T = 0$ implies $\lambda(X) = X$, our map λ keeps every *vector* of $V \cap V'$ fixed.

Let us start conversely with a linear map $\lambda : V \to V'$ which is onto and which keeps every vector of $V \cap V'$ fixed. Denote by T_0 the set of all vectors of the form $\lambda(X) - X$ where X ranges over V. It is obvious that T_0 is a subspace of Ω; its intersection with V is 0. Indeed, if $\lambda(X) - X \in V$, then $\lambda(X) \in X + V = V$; since $\lambda(X) \in V'$ we have $\lambda(X) \in V \cap V'$ and consequently, $\lambda(X) = X$ showing that our vector $\lambda(X) - X = 0$. Similarly we can prove $T_0 \cap V' = 0$.

We have $T_0 \subset V' + V$ and, therefore, $V + T_0 \subset V' + V$. If $\lambda(X)$ is any vector in V', then

$$\lambda(X) = (\lambda(X) - X) + X \in T_0 + V.$$

This shows $V + V' \subset T_0 + V$, and, therefore, $V + V' = T_0 + V$. Similarly $V' + T_0 = V + V'$; since $V \cap T_0 = V' \cap T_0 = 0$ we can write $V + V' = V \oplus T_0 = V' \oplus T_0$. Let T_1 be a subspace of Ω which is supplementary to $V + V'$. Set $T = T_0 \oplus T_1$. Then

$$V \oplus T_0 \oplus T_1 = V \oplus T = \Omega = V' \oplus T.$$

We may project from $\bar{T}$. Since $\lambda(X) - X \in T_0 \subset T$ we get $\langle X \rangle + T = \langle \lambda(X) \rangle + T$, and this shows that our λ will induce the projection from $\bar{T}$.

The condition that $\lambda(X) = X$ for all $X \in V \cap V'$ has no direct geometric meaning in the projective spaces $\bar{V}$ and $\overline{V'}$ where individual vectors of Ω do not appear. Let us see if we can replace it by a condition which has a geometric meaning. The projection will leave every line of $V \cap V'$, i.e., every "point" of $\bar{V} \cap \overline{V'}$ fixed. Suppose now that a collineation $\sigma : \bar{V} \to \overline{V'}$ is given which leaves every "point" of $\bar{V} \cap \overline{V'}$ fixed. Does it come from a projection? Let us assume $\dim(V \cap V') \geq 2$. Should $n = 2$, then this is only possible if $V = V'$; our σ must be identity and is clearly induced by the identity map λ, i.e., σ comes from a projection trivially. If $n \geq 3$, we may use the fundamental theorem of projective geometry. The collineation σ is induced by a semi-linear map λ and this map λ will keep every line of $V \cap V'$ fixed. For $X \in V \cap V'$ we will get $\lambda(X) = \alpha X$ and can prove that α is the same for all $X \in V \cap V'$ as in the fundamental theorem of projective geometry by comparing independent vectors of the space $V \cap V'$ (whose dimension is ≥ 2). We may now change the map λ by the factor α^{-1} and the new map λ will still induce σ but satisfy $\lambda(X) = X$ for all $X \varepsilon V \cap V'$. Denote by μ the automorphism of k to which λ belongs. For an $X \neq 0$ of $V \cap V'$ we have $\lambda(aX) = a^{\mu}\lambda(X) = a^{\mu}X$ on one hand and $\lambda(aX) = aX$ on the other. This proves that μ is the identity automorphism, i.e., that λ is an ordinary homomorphism of V onto V'. Since it keeps every vector of $V \cap V'$ fixed, our collineation is indeed a projection. We postpone for a while the discussion of the cases where $\dim(V \cap V') \leq 1$.

In classical synthetic projective geometry a collineation of two subspaces $\bar{V}$ and $\overline{V'}$ of Ω is called a projectivity if it can be obtained by a sequence of projections. Each of these projections (possibly onto other subspaces) is induced by a linear map and consequently also the projectivity of $\bar{V}$ onto $\overline{V'}$. Let us consider first the special case of a projectivity of $\bar{V}$ onto $\bar{V}$. It is induced by a linear map, therefore, by an element of $GL(V)$ and is consequently an element of $PGL(\bar{V})$. The question arises whether every element of $PGL(\bar{V})$ is a projectivity. Let H be any hyperplane of V. Since $N > n$ we can find a subspace V' of Ω which intersects V in the hyperplane H. We can find a linear map of V onto V' which is identity on H and carries a given vector $A \varepsilon V$ which is outside H onto a given vector B of V' which is outside H. This map will come from a projection. We can then project back onto V and move B onto any given vector A' of V which is outside H. The combined map is a projectivity of $\bar{V}$ onto $\bar{V}$ which is induced by a map $\lambda : V \rightarrow V$ such that λ is identity on H and such that $\lambda(A) = A'$. We can, for instance, achieve a "stretching" $\lambda(A) = aA$ of the vector A or we can also achieve $\lambda(A) = A + C$ where C is any vector of H. In the early part of Chapter IV we are going to see that every element of $GL(V)$ can be obtained by a succession of maps of this type (with varying hyperplanes H). If we anticipate this result then we can say that the projectivities of $\bar{V}$ onto $\bar{V}$ are merely the elements of $PGL(\bar{V})$. If V and V' are different spaces we may first project $\bar{V}$ onto $\overline{V'}$ in some fashion and follow it by a projectivity of $\overline{V'}$ onto $\overline{V'}$, hence by an arbitrary element of $PGL(\overline{V'})$. It is clear now that any linear map of V onto V' induces a projectivity of $\bar{V}$ onto $\overline{V'}$. This explains the dominant role played by the group $PGL(\bar{V})$ in the synthetic approach.

We return to the question which collineations of a space $\bar{V}$ onto a space $\overline{V'}$ are projections from a suitable center. One has certainly to assume that every "point" of $\bar{V} \cap \overline{V'}$ remains fixed and we have seen that this is enough if $\dim(V \cap V') \geq 2$. In order to dispose of the remaining cases we have to strengthen the assumption. We shall assume that the given collineation σ is a projectivity, i.e., that it is obtained by a sequence of projections. This implies the simplification that σ is induced by a linear map λ of V onto V' and, as we have seen, is equivalent with this statement.

If $V \cap V' = 0$, then no condition on λ was necessary to construct the center T of the projection. This case is, therefore, settled in the affirmative.

There remains the case where $\dim(V \cap V') = 1$, when $\bar{V}$ and $\overline{V'}$ intersect in a point. Let $\langle A \rangle = V \cap V'$. Since $\sigma\langle A \rangle = \langle A \rangle$ we must have $\lambda(A) = \alpha A$ with $\alpha \neq 0$. If, conversely, λ is a linear map of V onto V' for which $\lambda(A) = \alpha A$ (and α can have any non-zero value in k for a suitable λ) then λ will induce a projectivity σ of $\bar{V}$ onto $\overline{V'}$ which keeps $\langle A \rangle$ fixed. Once σ is given, one has only a little freedom left in the choice of λ. We may replace $\lambda(X)$ by $\lambda_1(X) = \lambda(\beta X)$ provided λ_1 is again a linear map. But this implies that β is taken from the center of k. We must now see whether we can achieve $\lambda_1(A) = A$ for a suitable choice of β. This leads to the equation $\beta\alpha = 1$ and we see that such a β can only be found if α is in the center of k. Since α might take on any value one has to assume that k is commutative. We gather our results in the language of projective geometry.

THEOREM 2.27. *Let $\bar{\Omega}$ be a projective space, $\bar{V}$ and $\overline{V'}$ proper subspaces of equal dimension and σ a collineation of $\bar{V}$ onto $\overline{V'}$ which leaves all points of $\bar{V} \cap \overline{V'}$ fixed. We can state:*

1) *If $\dim_p (\bar{V} \cap \overline{V'}) \geq 1$, then σ is a projection of $\bar{V}$ onto $\overline{V'}$ from a suitable center $\bar{T}$ (σ is a perspectivity).*

2) *If $\bar{V}$ and $\overline{V'}$ have an empty intersection and if σ is a projectivity, then σ is again a projection from a suitable center.*

3) *If $\bar{V}$ and $\overline{V'}$ intersect in a point and if σ is a projectivity, then it is always a projection if the field k is commutative. If k is not commutative, then this will not always be true.*

The most important special case of this theorem, when $\bar{\Omega}$ is a plane and $\bar{V}$, $\overline{V'}$ distinct lines leads to the third of our alternatives. A projectivity of the line $\bar{V}$ onto the line $\overline{V'}$ which keeps the point of intersection fixed is in general a projection only if the field is commutative. It depends on the configuration of Pappus.

The question whether every collineation of a space $\bar{V}$ onto itself is a projectivity has a quite different answer. One must assume $\dim_p \bar{V} \geq 2$ and now the problem reduces to $PGL = PS$; this is equivalent with $A(k) = I(k)$ by our diagram, every automorphism of k must be inner.

In the classical case $k = R$, the field of real numbers, both conditions are satisfied. The field R is commutative and identity is the only automorphism of R.

Our last theorem in this paragraph concerns the characterisation of the identity.

THEOREM 2.28. *Let $\bar{V}$ be a projective space over a commutative field k. Let σ be a projectivity of $\bar{V}$ onto $\bar{V}$ which leaves $n + 1$ points fixed ($n =$ dim V) and assume no n of these points lie on a hyperplane. We assert that σ is the identity. This theorem is false if k is not commutative.*

Proof: Let $\langle A_0 \rangle, \langle A_1 \rangle, \cdots, \langle A_n \rangle$ be the $n + 1$ points. The vectors $A_1, A_2, \cdots, A_n$ are a basis of V and if $A_0 = \sum_{i=1}^n \alpha_i A_i$ then $\alpha_i \neq 0$. These statements follow easily from the assumption about the points. If we replace A_i by $\alpha_i A_i$ we obtain the simpler formula $A_0 = \sum_{i=1}^n A_i$.

The projectivity σ is induced by a linear map λ and since σ leaves our points fixed, we must have $\lambda(A_i) = \beta_i A_i$ and consequently $\lambda(A_0) = \sum_{i=1}^n \beta_i A_i$. But $\lambda(A_0) = \beta_0 A_0$ whence $\beta_i = \beta_0$, all β_ν are equal. From $\lambda(A_i) = \beta A_i$ we deduce for any vector X that $\lambda(X) = \beta X$ since k is commutative. This was our contention. If k is not commutative select $\alpha, \beta \in k$ such that $\alpha\beta \neq \beta\alpha$. Let λ be the linear map which sends A_i into αA_i ($i = 1, \cdots, n$). The vector $A_0 = \sum_{i=1}^n A_i$ is mapped onto αA_0. The collineation σ which λ induces leaves the points $\langle A_i \rangle$ fixed. The point $\langle X \rangle = \langle A_1 + \beta A_2 \rangle$ is moved into $\langle \alpha A_1 + \beta\alpha A_2 \rangle = \langle A_1 + \alpha^{-1}\beta\alpha A_2 \rangle$ which is different from $\langle X \rangle$.

11. The projective plane

Consider an affine Desarguian plane and let k be its field. Let V be a three dimensional left vector space over k and W a plane of V. We shall construct a new Desarguian plane Π_W as follows:

The "points" of Π_W shall be those lines L of V which are *not* contained in W.

The "lines" of Π_W shall be those planes U of V which are $\neq W$. The incidence relation shall merely be $L \subset U$.

Let $W = \langle A_2, A_3 \rangle$ and $V = \langle A_1, A_2, A_3 \rangle$. If $L \not\subset W$, then L is spanned by a unique vector $A_1 + \xi A_2 + \eta A_3$ and we shall associate (ξ, η) as coordinates to L.

A "line" U will have an intersection $\langle \alpha A_2 + \beta A_3 \rangle$ with W. U can be spanned by $\alpha A_2 + \beta A_3$ and by some vector $A_1 + \gamma A_2 + \delta A_3$. Since $L \not\subset W$, a "point" L on U is spanned by a unique vector of the form $(A_1 + \gamma A_2 + \delta A_3) + t(\alpha A_2 + \beta A_3)$. The coordinates of L are, therefore,

$$(t\alpha + \gamma, t\beta + \delta) = t(\alpha, \beta) + (\gamma, \delta).$$

This is the equation of a line in parametric form and the only restriction is that $(\alpha, \beta) \neq (0, 0)$. We see that Π_W is coordinatized by the field k and that it has, therefore, the same structure as our original Desarguian plane.

The plane Π_W is obtained from the projective plane $\bar{V}$ by deleting the "line" W and all the "points" on W.

We thus have the following picture: If we take a projective plane $\bar{V}$ and eliminate one line and all the points on it, we get a Desarguian plane. All the Desarguian planes which can be obtained from $\bar{V}$ by this process have the "same" type of affine geometry. If two lines of $\bar{V}$ meet on a point of the deleted line W, then they do not meet in Π_W and are, therefore, parallel. Each deleted point is characterized by a

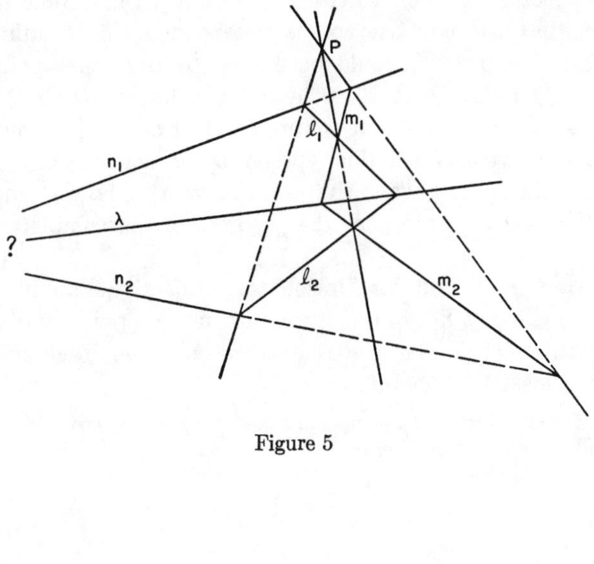

Figure 5

pencil of parallel lines of Π_W. We call them "infinite points" of Π_W and W itself the infinite line of Π_W.

From the affine configurations we can get projective configurations and prove them by deleting appropriate lines.

Figure 5[1] indicates a projective configuration. If we delete the line λ it becomes the affine Desargues: The lines l_1, l_2 and the lines m_1, m_2 become parallel so that the lines n_1, n_2 are also parallel; they must, therefore, meet in a point on the line λ. This configuration is known as the projective Desargues. It contains of course the affine Desargues as a special case if λ is the "infinite" line.

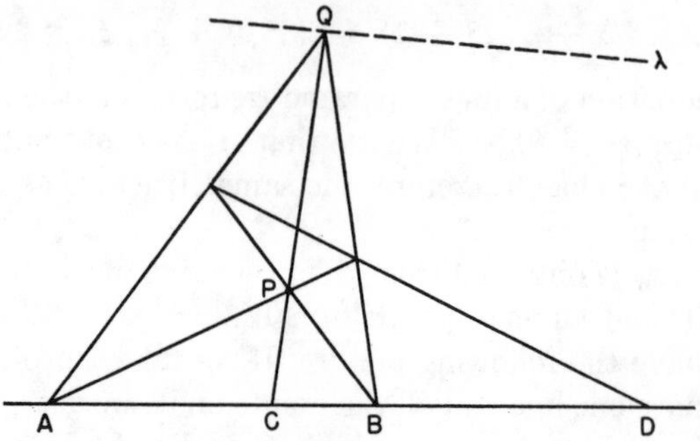

Figure 6

Figure 6 is another such configuration. Draw a line λ through the point Q which does not go through A, B, C. If we delete the line λ it becomes the construction of the fourth harmonic point D. The fact that the construction is independent in its affine form means that the point Q could be moved to any other point on λ, hence to any point Q' in $\bar{V}$ such that the line through Q and Q' does not go through A, B, or C (provided of course it is outside l). By two successive moves of this type we can bring it into any position. Thus the point D is always the same no matter how P and Q are chosen. This fact is known as the theorem of the complete quadrilateral.

The reader may find for himself the projective Pappus.

Any two distinct lines of $\bar{V}$ meet in a unique point and through distinct points of $\bar{V}$ there goes exactly one line. Each line of $\bar{V}$ contains at least three points.

[1]Figure 5 is meant to be in a plane. For a later purpose it is drawn in such a way that it could be interpreted in space.

It is now clear on what axioms one would base projective geometry in a plane:

1) Two distinct points determine a unique line on which they lie.

2) Two distinct lines intersect in exactly one point.

3) Each line contains at least 3 points and there are 3 non-collinear points.

4) The projective Desargues.

If $\bar{V}$ is a projective plane of this kind, then one deletes a line and all the points on it. One obtains an affine plane in which the affine Desargues holds. This affine plane can be coordinatized by a field k. With this field k one can construct a projective plane of the same structure as the given one. It is to be remarked, however, that the field is not canonically constructed, that it involves the choice of the deleted line. A canonical field associated with the plane would have to be explained as an equivalence class of fields (one for each deleted line).

We have never mentioned the axioms for affine or projective spaces of a dimension $n > 2$. It is approximately clear how these axioms would look. In the projective case the incidence axioms might be the rules connecting dimensions of intersection and sum, in the affine case the rules connecting the dimensions of intersection and join. We had given these rules in Chapter I. If $n > 2$ no Desargues is necessary, one can prove it from the incidence axioms. One look at Figure 5 suggests viewing the configuration as the projection of a three-dimensional configuration onto a plane. The three-dimensional configuration is easily proved. The reader should work out the details of such a proof. The theorem of Desargues is, therefore, the necessary and sufficient condition for a plane geometry to be extendable to a geometry in 3 (or more) dimensions. Indeed, one can prove Desargues if $n > 2$ and, on the other hand, coordinatize the plane if Desargues holds; but such a plane can clearly be imbedded in a higher dimensional space. This is again another geometric interpretation for Desargues.

If one replaces Desargues' theorem by Pappus, then, as Hessenberg has shown, Desargues can be implied from Pappus. It would be nice to have a proof of this fact in the style of our presentation.

One may also think of replacing Desargues' theorem by the theorem of the complete quadrilateral. Very interesting results have been obtained which the reader may look up in the literature.

EXERCISES. We can of course also associate a projective space $\bar{V}$ with a right vector space.

Let now V and V' be two vector spaces over fields k respectively k' (V can be right and V' either left or right, or V can be left and V' either left or right). Assume they have both the same dimension. Let $U \to U'$ be a one-to-one correspondence (onto) of the subspaces of V and the subspaces of V'. We shall call it

a *collineation* of $\bar{V}$ and $\overline{V'}$ if $U_1 \subset U_2$ implies $U_1' \subset U_2'$,

a *correlation* of $\bar{V}$ and $\overline{V'}$ if $U_1 \subset U_2$ implies $U_1' \supset U_2'$.

1) Let V be a left vector space over k. Find a right vector space V' over some field k' such that there exists a collineation of $\bar{V}$ onto $\overline{V'}$.

2) To each left vector space V there exists a canonically constructed right vector space V' over the same field k together with a canonical correlation $\bar{V} \to \overline{V'}$.

3) Use 1) and 2) to describe all collineations and correlations, making also use of the fundamental theorem of projective geometry.

4) Suppose there exists a correlation $\bar{V} \to \bar{V}$ (the same space). What is the necessary and sufficient condition on the field k of V? Consider the group of all correlations and collineations of $\bar{V} \to \bar{V}$ and investigate its structure. Draw diagrams.

Let τ be an antiautomorphism of k (if k is commutative every automorphism is also an antiautomorphism). Call a "product" $V \times V \to k$ a generalized pairing of V and V into k if the two distributive laws hold and if—in case V is a right vector space—the following conditions are satisfied:

$$X(Ya) = (XY)a, \qquad (Xa)Y = a^{\tau}(XY).$$

Making use of the dual $\hat{V}$ of V one can easily describe the laws governing such a pairing.

5) Suppose a correlation $\bar{V} \to \bar{V}$ exists. Prove that there exists a generalized pairing of V and V into k which allows us to describe the correlation algebraically. Find all pairings describing a given correlation.

6) Consider a correlation between two projective planes. Give the images of the configurations of Desargues, Pappus and the complete quadrilateral. Does this constitute a proof of the image configurations?

7) Let Π and Π' be two Desarguian affine planes, k respectively k' the fields of trace-preserving homomorphisms of the translation groups T respectively T'. Let $\lambda : \Pi \to \Pi'$ be a one-to-one map of Π onto Π' which preserves lines. Prove:

a) $\lambda T \lambda^{-1} = T'$.

b) For each $\alpha \in k$ the map

$$\tau' \to \lambda(\lambda^{-1}\tau'\lambda)^{\alpha}\lambda^{-1}$$

is a trace-preserving homomorphism of T', hence a certain element of k' which we may denote by $\alpha^{\bar{\lambda}}$.

c) The map $\bar{\lambda} : k \to k'$ given by $\alpha \to \alpha^{\bar{\lambda}}$ is an isomorphism of k onto k'.

d) If coordinates are given in Π then the coordinate system of Π' can be placed in such a position that the image of the point (x, y) in Π is the point $(x^{\bar{\lambda}}, y^{\bar{\lambda}})$ of Π'.

e) Describe all such maps λ by formulas.

8) Using Figure 6 give a sequence of three projections which thereby produce a projectivity σ of the line l through A, B onto itself with the following properties: A and B are fixed points; if C is distinct from A and B then its image under σ is the fourth harmonic point D. What is the order of the projectivity σ?

9) Let k be a commutative field of characteristic $\neq 2$, V a vector space over k and $\bar{V}$ the corresponding projective space. Let σ be a projectivity of $\bar{V}$ onto itself which is of order 2 and has at least one fixed "point". Prove that among the linear maps of V onto itself which induce σ there is one—call it λ—which is also of order 2. Show that V is the direct sum of two subspaces U and W such that λ keeps every vector of U fixed and reverses every vector of W. The projectivity σ is completely described by the pair U, W of subspaces (U, W equivalent to W, U).

10) Let k be commutative and of characteristic $\neq 2$. Show that a projectivity of a line l which is of order 2 and which has a fixed point is by necessity a map of the type described in 8). Avoid any computation. Returning to 9) describe the projectivity σ by geometric constructions. Assume that k is the field of quaternions. Give a projectivity of a line which is of order 2 and is not of this type.

CHAPTER III

Symplectic and Orthogonal Geometry

1. Metric structures on vector spaces

This chapter presupposes a knowledge of Chapter I, §§ 2–4. We will deal exclusively with vector spaces V of finite dimension n over a *commutative* field k. A left space V shall be made canonically into a two-sided space by the definition

$$Xa = aX , \qquad a \in k, \ X \in V.$$

Since V has become a left space as well as a right space we can consider pairings of V and V into k. In such a pairing a product $XY \in k$ is defined for all $X, Y \in V$ and will now satisfy the rules

$$(3.1) \quad \begin{aligned} (X_1 + X_2)Y &= X_1Y + X_2Y, & X(Y_1 + Y_2) &= XY_1 + XY_2 , \\ (aX)Y &= a(XY), & X(aY) &= a(XY) \end{aligned}$$

(the last rule since $aY = Ya$ and since k is commutative).

The reader should think intuitively of $X^2 = XX$ as something like "length" of the vector X and of XY as something related to the "angle" between X and Y. We shall say that such a product defines a "metric structure" on V and investigate first how such a structure can be described in terms of a basis $A_1 , A_2 , \cdots , A_n$ of V.

We set

$$(3.2) \qquad A_iA_j = g_{ij} \in k, \qquad i, j = 1, 2, \cdots , n.$$

Let

$$(3.3) \qquad X = \sum_{\nu=1}^{n} x_\nu A_\nu , \qquad Y = \sum_{\mu=1}^{n} y_\mu A_\mu$$

be any two vectors. The rules (3.1) allow us to express their product:

$$(3.4) \qquad XY = \sum_{\nu,\mu=1}^{n} g_{\nu\mu}x_\nu y_\mu ,$$

which shows that we know XY if the g_{ij} are known.

Let us select conversely arbitrary elements g_{ij} in k and define a product XY (of vectors X and Y given by (3.3)) by means of (3.4). This product satisfies obviously the rules (3.1) and is, therefore, a pairing. If, in (3.4), we specialise X to A_i and Y to A_j we again get (3.2). A function of the variables x_ν and y_μ of the type (3.4) is called bilinear. Hence the study oı bilinear functions is equivalent with a study of metric structures on V.

For a given pairing the g_{ij} will depend on the choice of the basis. Let us see how a change of the basis will affect the g_{ij}. Let $B_1, B_2, \cdots, B_n$ be a new basis of V; then $B_j = \sum_{\nu=1}^{n} A_\nu a_{\nu j}$ with certain $a_{ij} \varepsilon k$. The B_j will form a basis of V if and only if the determinant of the matrix (a_{ij}) is $\neq 0$. With the new basis we construct the new $\overline{g_{ij}}$:

$$\overline{g_{ij}} = B_i B_j = \sum_{\nu,\mu=1}^{n} A_\nu a_{\nu i} A_\mu a_{\mu j} = \sum_{\nu,\mu=1}^{n} a_{\nu i} g_{\nu\mu} a_{\mu j} .$$

In matrix form this can be written as

$$(3.5) \qquad (\overline{g_{ij}}) = (a_{ji})(g_{ij})(a_{ij})$$

where (a_{ji}) is the transpose of the matrix (a_{ij}).

Our pairing will have a left kernel U_0 and a right kernel V_0. They are now of course subspaces of the same space V. We know $\dim V/U_0 = \dim V/V_0$ which implies that U_0 and V_0 have the same dimension. Let us compute U_0 in terms of our basis A_ν. A vector $X = \sum_{\nu=1}^{n} x_\nu A_\nu$ will be in U_0 if and only if $XY = 0$ for all Y. Then certainly $XA_j = 0$ $(j = 1, 2, \cdots, n)$. Conversely, if $XA_j = 0$ for $j = 1, 2, \cdots, n$ we get $X \cdot (\sum_{\mu=1}^{n} y_\mu A_\mu) = 0$. X will be in U_0 if and only if

$$(3.6) \qquad \sum_{\nu=1}^{n} g_{\nu j} x_\nu = 0, \qquad j = 1, 2, \cdots, n.$$

We are especially interested in the question: when will U_0 be merely the 0 subspace of V? Since V_0 has the same dimension this also implies that $V_0 = 0$. The equations (3.6) should only have the trivial solution and this is the case if and only if the determinant of the matrix (g_{ij}) is different from 0. We shall use a certain terminology.

Definition 3.1. We call a vector space V with a metric structure non-singular if the kernels of the pairing are 0. The determinant

$G = \det(g_{ij})$ shall be called the discriminant of V. The necessary and sufficient condition for V to be non-singular is $G \neq 0$.

Let us return to (3.5) and denote by $\bar{G}$ the determinant of $(\overline{g_{ij}})$ (hence the discriminant of V as defined by the basis B_ν). Taking determinants in (3.5) we obtain

$$\bar{G} = G \cdot (\det (a_{ij}))^2. \tag{3.7}$$

The geometric set-up alone determines, therefore, the discriminant of V only up to a square factor.

DEFINITION 3.2. Let V and W be vector spaces over k, and σ: $V \rightarrow W$ be an *isomorphism* of V into W. Suppose that metric structures are defined on both V and W. We shall call σ an isometry of V into W if σ "preserves products":

$$XY = (\sigma X)(\sigma Y) \quad \text{for all} \quad X, Y \in V.$$

The most important case is that of isometries of V onto V. Then σ^{-1} is also such an isometry and if σ and τ are isometries of V onto V so is $\sigma\tau$.

DEFINITION 3.3. Let V be a space with a metric structure. The isometries of V onto V form a group which we will call the group of V.

Let σ be an endomorphism of V into V and A_ν a basis of V. Put $\sigma A_i = B_i$; if σ is an isometry, then we must have $B_iB_j = A_iA_j$ $(i, j = 1, 2, \cdots, n)$. Suppose, conversely, that these equations hold. If $X = \sum_{\nu=1}^{n} x_\nu A_\nu$ and $Y = \sum_{\mu=1}^{n} y_\mu A_\mu$ then

$$\sigma X = \sum_{\nu=1}^{n} x_\nu(\sigma A_\nu) = \sum_{\nu=1}^{n} x_\nu B_\nu$$

and

$$\sigma Y = \sum_{\mu=1}^{n} y_\mu B_\mu .$$

We have

$$XY = \sum_{\nu,\mu=1}^{n} x_\nu y_\mu A_\nu A_\mu = \sum_{\nu,\mu=1}^{n} x_\nu y_\mu B_\nu B_\mu = \sigma X \cdot \sigma Y.$$

The map σ will be an isometry if we can show (or assume if necessary) that the kernel of σ is 0. If V is non-singular this follows automatically. Indeed, if $\sigma X = 0$, then certainly $\sigma X \cdot \sigma Y = 0$ for all $Y \varepsilon V$. But $\sigma X \sigma Y = XY$ so that $XY = 0$ for all $Y \varepsilon V$. This means that X is in the left kernel U_0 and therefore 0, if we assume V to be non-singular.

THEOREM 3.1. *Let σ be an endomorphism $V \to V$, A_i a basis of V and $\sigma A_i = B_i$; σ will be an isometry if and only if $A_iA_j = B_iB_j$ for all $i, j = 1, 2, \cdots, n$ and if the kernel of σ is 0. This last condition is unnecessary if V is non-singular.*

If we write $B_i = \sum_{\nu=1}^{n} A_\nu a_{\nu i}$, then the matrix (a_{ij}) is the one describing the endomorphism σ in the sense of Chapter I, §3. Since $B_iB_j = \overline{g_{ij}} = g_{ij}$ for an isometry, we get from (3.5) and (3.7)

$$(g_{ij}) = (a_{ji})(g_{ij})(a_{ij}), \tag{3.8}$$

$$G = (\det(a_{ij}))^2 \cdot G = (\det \sigma)^2 \cdot G. \tag{3.9}$$

Should V be non-singular, then $G \neq 0$ and we get $\det \sigma = \pm 1$.

THEOREM 3.2. *If V is non-singular and σ an isometry of V onto V, then* $\det \sigma = \pm 1$. *If* $\det \sigma = +1$, *then σ is called a rotation; if* $\det \sigma = -1$, *then we call σ a reflexion. The rotations form an invariant subgroup of the group of V whose index is at most* 2.

The last part follows from the fact that the map $\sigma \to \det \sigma$ is a homomorphism of the group of V whose kernel are the rotations and whose image is either ± 1 or just $+1$ (in case there are no reflections or if the characteristic of k is 2).

If we form $X^2 = X \cdot X$ we find from (3.4)

$$X^2 = \sum_{i,j=1}^{n} g_{ij} x_i x_j . \tag{3.10}$$

This is an expression which depends quadratically on the x_ν and is, therefore, called a quadratic form. The coefficient of x_i^2 is g_{ii}, the coefficient of x_ix_j is $g_{ij} + g_{ji}$ if $i \neq j$. We notice that we can select the g_{ij} in such a way that X^2 becomes a given quadratic form; it is even possible in several ways.

One can define a quadratic form intrinsically (without reference to a basis) as a map Q of V into k (but not a linear one) which satisfies the two conditions:

1) $Q(aX) = a^2Q(X), \quad a \varepsilon k, \quad X \varepsilon V.$

2) The function of two variables $X, Y \varepsilon V$ given by

$$Q(X + Y) - Q(X) - Q(Y),$$

which we shall denote by $X \circ Y$, is a pairing of V and V into k.

Indeed, if we start with an arbitrary pairing XY and define $Q(X) = X^2$, then this $Q(X)$ satisfies condition 1). For $X \circ Y$ one finds

$$X \circ Y = (X + Y)^2 - X^2 - Y^2 = XY + YX.$$

It will satisfy condition 2). We remark, however, that $X \circ Y$ is not the original pairing.

Suppose now, conversely, that $Q(X)$ satisfies the two conditions. If we put $X = X_1 + X_2 + \cdots + X_{r-1}$ and $Y = X_r$ in condition 2) we find

$$Q(X_1 + X_2 + \cdots + X_r) = Q(X_1 + X_2 + \cdots + X_{r-1}) + Q(X_r) + \sum_{i=1}^{r-1} (X_i \circ X_r).$$

If we use induction on r we get

$$Q(X_1 + X_2 + \cdots + X_r) = \sum_{i=1}^{r} Q(X_i) + \sum_{1 \leq i < j \leq r} (X_i \circ X_j).$$

Let $A_1, A_2, \cdots, A_n$ be a basis of V and let $X = \sum_{i=1}^{n} x_i A_i$. Then

$$Q(X) = \sum_{i=1}^{n} x_i^2 Q(A_i) + \sum_{1 \leq i < j \leq n} x_i x_j (A_i \circ A_j).$$

This shows that $Q(X)$ depends indeed quadratically on the x_i.

The pairing

(3.11) $$X \circ Y = Q(X + Y) - Q(X) - Q(Y),$$

which we get from a quadratic form, is very special. If we put $Y = X$ in (3.11) and notice that $Q(2X) = 4Q(X)$ we obtain

(3.12) $$X \circ X = 2Q(X).$$

Equation (3.11) shows also that

(3.13) $$X \circ Y = Y \circ X.$$

We distinguish now two cases:

a) The characteristic of k is $\neq 2$. Then we can write $Q(X) = \frac{1}{2}(X \circ X)$ and see that the quadratic form differs inessentially from

$X \circ X$. The pairing $X \circ Y$ is symmetric. Can there be another symmetric pairing $XY = YX$ for which we also have $Q(X) = \frac{1}{2}X \cdot X$? We would then get

$$Q(X + Y) - Q(X) - Q(Y)$$
$$= \tfrac{1}{2}(X + Y)(X + Y) - \tfrac{1}{2}XX - \tfrac{1}{2}YY = X \cdot Y$$

which shows $X \cdot Y = X \circ Y$.

The quadratic form $Q(X)$ determines, therefore, uniquely a *symmetric* pairing such that

$$X \cdot X = 2Q(X).$$

If our aim is the study of quadratic forms we may just as well start with a symmetric pairing and call X^2 the corresponding quadratic form since it differs from $Q(X)$ by the inessential constant factor 2.

b) The situation is very different if the characteristic of k is 2. Equation (3.12) becomes $X \circ X = 0$. If we start with a symmetric pairing $XY = YX$ and hope to obtain $Q(X)$ from X^2 we do not succeed. Indeed, if $X = \sum_{i=1}^{n} A_i x_i$, then

$$X^2 = \sum_{i=1}^{n} A_i^2 \cdot x_i^2$$

since the two terms $A_i A_j x_i x_j$ and $A_j A_i x_j x_i$ (for $i \neq j$) will cancel if the characteristic is two. One can for instance never obtain a term $x_1 x_2$. To start with an unsymmetric pairing is not a desirable procedure since such a pairing is not uniquely determined by the quadratic form.

2. *Definitions of symplectic and orthogonal geometry*

Suppose again that an arbitrary pairing of V and V into k is given. As in any pairing we shall call a vector A orthogonal to a vector B if $AB = 0$. This raises immediately the problem: In which metric structures does $AB = 0$ imply that $BA = 0$? Let us suppose that V has this property and let $A, B, C \in V$. Remembering that k is commutative we get

$$A((AC)B - (AB)C) = (AC)(AB) - (AB)(AC) = 0.$$

Hence

$$((AC)B - (AB)C)A = 0$$

or

$$(3.14) \qquad (AC)(BA) = (CA)(AB).$$

For $C = A$ we obtain $A^2 \cdot (BA) = A^2 \cdot (AB)$. Should $A^2 \neq 0$ then we can conclude $BA = AB$. In other words:

If $AB \neq BA$ then $A^2 = 0$ and similarly $B^2 = 0$.

Suppose now that V contains two special vectors A and B such that $AB \neq BA$. We intend to show that $C^2 = 0$ for any vector C. This would certainly be true if $AC \neq CA$ so that we may assume $AC = CA$. But (3.14) is then consistent with $AB \neq BA$ only if $AC = CA = 0$. We may interchange A and B and see that we can also assume $BC = CB = 0$. But now

$$(A + C)B = AB \neq BA = B(A + C).$$

Consequently $(A + C)^2 = 0$; since $A^2 = 0$ (from $AB \neq BA$) and $AC = CA = 0$ we get indeed $C^2 = 0$. We see that there are two types of metric structures with our property:

1) The *symplectic geometry.*

Here we postulate

$$(3.15) \qquad X^2 = 0 \quad \text{for all} \quad X \in V.$$

Replacing the vector X by $X + Y$ we get $X^2 + XY + YX + Y^2 = 0$, hence

$$(3.16) \qquad XY = -YX, \qquad X, Y \in V.$$

Equation (3.16) shows that $AB = 0$ implies indeed $BA = 0$.

From condition (3.16) one can not get (3.15) as the special case $Y = X$, since k may have characteristic 2.

For the g_{ij} we obtain in a symplectic geometry

$$(3.17) \qquad g_{ii} = 0, \qquad g_{ij} = -g_{ji}\,.$$

A bilinear form $\sum_{i,j=1}^{n} g_{ij}x_iy_j$ for which (3.17) holds is called skew symmetric. If (3.17) is satisfied, $X^2 = \sum_{i,j=1}^{n} g_{ij}x_ix_j = 0$ which shows that (3.17) is the necessary and sufficient condition for a symplectic geometry and that such a geometry is equivalent with the study of skew symmetric bilinear forms.

2) The *orthogonal geometry.*

If V is not symplectic but if it has our property, then, by necessity,

$$(3.18) \qquad XY = YX \quad \text{for all} \quad X, Y \in V.$$

This is a *symmetric* pairing which we have also met a while ago in connection with quadratic forms. In this case $AB = 0$ implies again $BA = 0$.

If the characteristic of k is $\neq 2$, this geometry is entirely satisfactory since the symmetric pairings are in one-to-one correspondence with quadratic forms and one may simply say that X^2 is the quadratic form connected with our pairing.

If, however, k has characteristic 2, then the symmetric pairings are not general enough. In this case one *starts* with any quadratic form $Q(X)$ and defines a pairing by (3.11). But let us simplify the notation and write this pairing as XY. Hence

$$(3.19) \qquad XY = Q(X + Y) - Q(X) - Q(Y).$$

Equation (3.12) shows that

$$(3.20) \qquad X^2 = 0.$$

The underlying pairing is, therefore, symplectic (if the characteristic is 2, then there is no difference between (3.18) and (3.16)). In intuitive geometric language: V has a symplectic geometry refined by an additional quadratic form (measuring lengths if you wish) which is related to the symplectic geometry by (3.19). The geometry of V is called in both cases an *orthogonal geometry.*

To repeat the connections with the quadratic form: $(X + Y)^2 = X^2 + Y^2 + 2XY$ if the characteristic $\neq 2$, $Q(X + Y) = Q(X) + Q(Y) + XY$ for characteristic 2. One may call these two formulas the Pythagorean theorem or also the "law of cosine" which they become in special cases.

But we have to restrict ourselves. For the sake of simplicity we shall *always assume* tacitly that the field has a characteristic $\neq 2$ if we deal with an *orthogonal* geometry. In the case of a symplectic geometry we shall not make any restrictions.

The interested reader may consult the quoted literature for more details on the case of an orthogonal geometry when the characteristic is 2.

Appendix for readers who have worked on some exercises in Chapter II, §11.

If k is not necessarily commutative but has an antiautomorphism τ, we may consider a generalized pairing of a right vector space

V with itself into k and investigate the corresponding problem: Suppose $XY = 0$ implies always $YX = 0$. Since the antiautomorphism and the pairing are involved the discussion is a little harder. Assume $\dim V \geq 2$ and consider first the case where the kernel of the pairing is 0.

Let $A, B, C \in V$ such that $AB \neq 0$ and $AC \neq 0$. From

$$A(B \cdot (AB)^{-1} - C \cdot (AC)^{-1}) = 0$$

we get

$$(B \cdot (AB)^{-1} - C \cdot (AC)^{-1})A = 0$$

which leads to

$$(AB)^{-\tau}(BA) = (AC)^{-\tau}(CA) = \alpha.$$

We may write $(BA) = (AB)^{\tau}\alpha$ where α depends only on A; for vectors B for which $AB = BA = 0$ this equation is trivially true. One has now to show that α is independent of A. Let A_1 and A_2 be two independent vectors. There exists a vector B_1 which is orthogonal to A_2 but not to A_1 ; i.e., $A_1B_1 \neq 0$ and $A_2B_1 = 0$. Changing B_1 by a factor we may assume $A_1B_1 = 1$, $A_2B_1 = 0$. There exists also a vector B_2 such that $A_1B_2 = 0$, $A_2B_2 = 1$. The vector $B = B_1 + B_2$ satisfies $A_1B = 1$, $A_2B = 1$. We know three equations:

$$BA_1 = (A_1B)^{\tau}\alpha_1 , \qquad BA_2 = (A_2B)^{\tau}\alpha_2 ,$$

$$B(A_1 - A_2) = ((A_1 - A_2)B)^{\tau}\alpha_3 .$$

They give $BA_1 = \alpha_1$, $BA_2 = \alpha_2$, $\alpha_1 - \alpha_2 = 0$. Should A_1 and A_2 be dependent vectors $\neq 0$, then we compare them to a third independent vector and conclude that α is the same for all $A \neq 0$. For $A = 0$ it does not matter what α we take. The result is:

$$(YX) = (XY)^{\tau}\alpha \quad \text{for all} \quad X, Y \in V.$$

Case I. $X^2 = 0$ for all $X \in V$. We deduce then, as in the symplectic case, $YX = -(XY)$ and obtain $-(XY) = (XY)^{\tau}\alpha$. Selecting X and Y in such a way that $XY = 1$ we obtain $\alpha = -1$, and are left with $(XY) = (XY)^{\tau}$. But XY can range over all of k and we see that $\tau = 1$, the identity. The identity can be an antiautomorphism only if the field k is commutative. This means that our geometry is symplectic.

Case II. There exists a special vector X_0 such that $X_0^2 = \beta \neq 0$.

Set $X = Y = X_0$ in our equation and we obtain $\alpha = \beta^{-\tau}\beta$. Let us introduce a new pairing $[XY] = \beta^{-1}(XY)$. It differs very inessentially from the given pairing (just by a constant factor). We get

$$\beta[YX] = (\beta[XY])^{\tau}\beta^{-\tau}\beta = [XY]^{\tau}\beta,$$

$$[YX] = \beta^{-1}[XY]^{\tau}\beta.$$

We have

$$[(Ya)X] = \beta^{-1}((Ya)X) = \beta^{-1}a^{\tau}(YX) = \beta^{-1}a^{\tau}\beta\cdot[YX]$$

which shows that the new pairing belongs to the antiautomorphism $a \to \beta^{-1}a^{\tau}\beta = a^{\lambda}$. With λ we can now write

$$[YX] = [XY]^{\lambda}.$$

Interchanging once more: $[YX] = [YX]^{\lambda^2}$ which implies that $\lambda^2 = 1$, the identity.

Making this replacement in our pairing we may, therefore, assume that τ is an antiautomorphism of order at most 2 and that

$$(YX) = (XY)^{\tau}.$$

If $\tau = 1$, then the field must be commutative and we are in the case of orthogonal geometry.

If τ is of order 2, then k may or may not be commutative. The corresponding geometry is called *unitary* and the form X^2 is called a *hermitian form.*

If the pairing has a kernel $V_0 \neq 0$, one has to assume dim $V/V_0 \geq 2$ and obtains the same result by going to the pairing induced on V/V_0.

EXERCISE. Each of our geometries with kernel 0 induces on the projective space $\bar{V}$ a correlation of order 2 and each correlation of order 2 is induced by one of our geometries. By what geometric feature is the symplectic correlation distinguished from the others? Is the name "symplectic" somewhat justified? Suppose the geometry is either unitary or orthogonal. What does it mean for a special vector to satisfy $X^2 = 0$ in terms of the correlation? This leads to the classical definition of a conic in projective geometry.

3. *Common features of orthogonal and symplectic geometry*

In this paragraph we will study spaces with either an orthogonal or a symplectic geometry. In either case orthogonality of vectors or of subspaces is unambiguously defined. If U is a subspace of V, then the orthogonal subspace U^* has a unique meaning. The two kernels of the pairing are the same, they are the space V^*.

DEFINITION 3.4. The kernel V^* of V is called the radical of V and denoted by rad V.

If U is a subspace of V, then the pairing of V induces by restriction a pairing of U which will be of the same type as the one of V— orthogonal or sympletic. U itself has a radical consisting of those vectors of U^* which are in U. In other words

$$\text{rad } U = U \cap U^*, \qquad U \subset V. \tag{3.21}$$

DEFINITION 3.5. If V is the direct sum

$$V = U_1 \oplus U_2 \oplus \cdots \oplus U_r \tag{3.22}$$

of subspaces which are mutually orthogonal, then we shall say that V is the orthogonal sum of the U_i and use the symbol

$$V = U_1 \perp U_2 \perp \cdots \perp U_r\,. \tag{3.23}$$

Let V be a vector space which is a direct sum (3.22) of subspaces U_i . Suppose that a geometric structure is given to each subspace. Then there is a unique way to extend these structures to one of V such that V becomes the orthogonal sum of the U_i .

If

$$X = \sum_{i=1}^{n} A_i\,, \qquad Y = \sum_{i=1}^{n} B_i$$

are vectors of V and A_i , $B_i \in U_i$, then we have obviously to define

$$XY = A_1B_1 + A_2B_2 + \cdots + A_rB_r\,. \tag{3.24}$$

The reader will have no difficulty in proving that (3.24) defines a pairing on V and that V will have a symplectic, respectively, orthogonal geometry if all the U_i are symplectic, respectively, orthogonal. The geometry of V induces on each U_i its original geometry and U_i and U_j are orthogonal if $i \neq j$.

Suppose $V = U_1 + U_2 + \cdots + U_r$, U_i orthogonal to U_j if $i \neq j$, but do not assume that the sum is direct. Let $X = \sum_{i=1}^{n} A_i$, $A_i \in U_i$, and assume $X \in \text{rad } V$. Then we must have $XB_i = 0$ for all $B_i \in U_i$ which gives $A_iB_i = 0$ or $A_i \in \text{rad } U_i$. Conversely, if each

$A_i \in \text{rad } U_i$, then $X \in \text{rad } V$. In other words, if the U_i are mutually orthogonal, then,

$$\text{rad } V = \text{rad } U_1 + \text{rad } U_2 + \cdots + \text{rad } U_r . \tag{3.25}$$

Should each U_i be non-singular, i.e., $\text{rad } U_i = 0$, then we obtain $\text{rad } V = 0$, V is non-singular. But in this case our sum is direct. Indeed if $\sum_{i=1}^{n} A_i = 0$ we obtain $A_i B_i = 0$ for any $B_i \in U_i$. Hence $A_i \in \text{rad } U_i = 0$. We can, therefore, write (3.23) if each U_i is non-singular.

Consider the subspace rad V of V and let U be a supplementary subspace

$$V = \text{rad } V \oplus U;$$

rad V is orthogonal to V and, therefore, to U and we get

$$V = \text{rad } V \perp U. \tag{3.26}$$

We deduce

$$\text{rad } V = \text{rad}(\text{rad } V) \perp \text{rad } U = \text{rad } V \perp \text{rad } U.$$

Since this last sum is direct we must have $\text{rad } U = 0$. U is, therefore, non-singular.

The geometry on V does not in general induce a geometry on a factor space. It does so, however, for the space $V/\text{rad } V$. It is natural to define as product of the coset$(X + \text{rad } V)$ and the coset$(Y + \text{rad } V)$ the element XY:

$$(X + \text{rad } V)\cdot(Y + \text{rad } V) = XY. \tag{3.27}$$

In Theorem 1.3 we have mapped a supplementary space U isomorphically onto the factor space $V/\text{rad } V$. The map was canonical and sent the vector X of U onto the coset $X + \text{rad } V$. The equation (3.27) means that this map is an isometry of U onto $V/\text{rad } V$. Thus we have proved

THEOREM 3.3. *Each space U which is supplementary to* rad V *gives rise to an orthogonal splitting* (3.26). *U is non-singular and canonically isometric to* $V/\text{rad } V$.

DEFINITION 3.6. Let $V = U_1 \perp U_2 \perp \cdots \perp U_r$, $V' = U_1' \perp U_2' \perp \cdots \perp U_r'$ be orthogonal splittings of two spaces V and V' and suppose that an isometry σ_i of U_i into U_i' is given for each i. If

$X = \sum_{i=1}^{n} A_i$ with $A_i \in U_i$ is a vector of V, then we can define a map σ of V into V' by

$$\sigma X = \sigma_1 A_1 + \sigma_2 A_2 + \cdots + \sigma_r A_r \tag{3.29}$$

which is an isometry and shall be denoted by

$$\sigma = \sigma_1 \perp \sigma_2 \perp \cdots \perp \sigma_r . \tag{3.30}$$

We shall call it the orthogonal sum of the maps σ_i .

The reader will, of course, have to check that σ is an isomorphism of V into V' and that scalar products are preserved.

THEOREM 3.4. *Let $V = U_1 \perp U_2 \perp \cdots \perp U_r$ and each σ_i an isometry U_i onto U_i . Then the orthogonal sum*

$$\sigma = \sigma_1 \perp \sigma_2 \perp \cdots \perp \sigma_r$$

is an isometry of V onto V and we have

$$\det \sigma = \det \sigma_1 \cdot \det \sigma_2 \cdot \cdots \cdot \det \sigma_r . \tag{3.31}$$

If

$$\tau = \tau_1 \perp \tau_2 \perp \cdots \perp \tau_r$$

where the τ_i are also isometries of U_i onto U_i then

$$\sigma\tau = \sigma_1\tau_1 \perp \sigma_2\tau_2 \perp \cdots \perp \sigma_r\tau_r . \tag{3.32}$$

The proof of this theorem is straight forward and can be left to the reader.

THEOREM 3.5. *Suppose that V is non-singular and let U be any subspace of V. We have always*

$$U^{**} = U, \qquad \dim U + \dim U^* = \dim V, \tag{3.33}$$

$$\operatorname{rad} U = \operatorname{rad} U^* = U \cap U^*. \tag{3.34}$$

The subspace U will be non-singular if and only if U^* is non-singular. Should U be non-singular, then we have

$$V = U \perp U^*. \tag{3.35}$$

Finally, if $V = U \perp W$, then U and W are non-singular and $W = U^*$.

Proof: Since the kernels of our pairing are 0, we have (3.33) from the general theory of pairings. Formulas (3.33) and (3.21) imply (3.34). If U is non-singular, then (3.34) shows that U^* is non-singular and that the sum $U + U^*$ is direct; since the dimensions fit, we get (3.35). If $V = U \perp W$, then $W \subset U^*$ and $\dim W = n - \dim U = \dim U^*$; therefore $W = U^*$ and $\operatorname{rad} U = U \cap U^* = 0$.

DEFINITION 3.7. A space is called isotropic if all products between vectors of the space are 0. The zero subspace of a space and the radical of a space are examples of isotropic subspaces. A vector A is called isotropic if $A^2 = 0$. In a symplectic geometry every vector is isotropic.

THEOREM 3.6. *Let V be a space with orthogonal geometry and suppose that every vector of V is isotropic. Then V is isotropic.*

Proof: Under our assumption this geometry is symplectic as well as orthogonal. We have, therefore, $XY = -YX = YX$. This implies $XY = 0$ since we assume, in case of an orthogonal geometry, that the characteristic of k is $\neq 2$.

The following special case plays an important role in the general theory.

We assume that $\dim V = 2$, that V is non-singular, but that V contains an isotropic vector $N \neq 0$.

If A is any vector which is not contained in the line $\langle N \rangle$, then $V = \langle N, A \rangle$. We shall try to determine another isotropic vector M such that $NM = 1$. Put $M = xN + yA$, then $NM = yNA$. If NA were 0, then $N \in \operatorname{rad} V$; but we have assumed V to be non-singular. Therefore $NA \neq 0$ and we can determine y uniquely so that $NM = 1$. In the symplectic case $M^2 = 0$ is automatically satisfied so that any x is possible. If V has orthogonal geometry, x must be determined from

$$M^2 = 0 = 2xyNA + y^2A^2.$$

This is also possible since $2yNA \neq 0$ and leads to a uniquely determined x.

We have now for both geometries

$$V = \langle N, M \rangle, \qquad N^2 = M^2 = 0, \qquad NM = 1.$$

Conversely, if $V = \langle N, M \rangle$ is a plane, we may impose on it a geometry by setting $g_{11} = g_{22} = 0$ and $g_{12} = 1$ (hence $g_{21} = 1$ in the orthogonal, $g_{21} = -1$ in the symplectic case). Then $N^2 = M^2 = 0$ and $NM = 1$. If $X = xN + yM \in \operatorname{rad} V$, then $XM = 0$; hence $x = 0$ and $NX = 0$ which gives $y = 0$. V is non-singular.

Suppose $V = \langle N, M\rangle$ has an orthogonal geometry. $X = xN + yM$ will be isotropic if $X^2 = 2xy = 0$; hence either $y = 0$, $X = xN$ or $x = 0$, $X = yM$.

DEFINITION 3.8. A non-singular plane which contains an isotropic vector shall be called a *hyperbolic plane*. It can always be spanned by a pair N, M of vectors which satisfy

$$N^2 = M^2 = 0, \qquad NM = 1.$$

We shall call any such *ordered* pair N, M a hyperbolic pair. If V is a non-singular plane with orthogonal geometry and $N \neq 0$ is an isotropic vector of V, then there exists precisely one M in V such that N, M is a hyperbolic pair. The vectors xN and yM are then the only isotropic vectors of V.

DEFINITION 3.9. An orthogonal sum of hyperbolic planes $P_1, P_2, \cdots, P_r$ shall be called a hyperbolic space:

$$H_{2r} = P_1 \perp P_2 \perp \cdots \perp P_r .$$

It is non-singular and of course of even dimension $2r$.

We may call a space irreducible if it can not be written as an orthogonal sum of proper subspaces. Because of (3.26) we see that an irreducible space is necessarily either non-singular or isotropic. If it is isotropic, then its dimension is 1 since any direct splitting of an isotropic space is also an orthogonal splitting. To discuss the non-singular case we distinguish:

1) *Orthogonal geometry.* Because of Theorem 3.6, V must contain a non-isotropic vector A. The subspace $U = \langle A\rangle$ is non-singular and (3.35) shows $U^* = 0$, $\dim V = 1$.

2) *Symplectic geometry.* Let $N \neq 0$ be any vector of V. Since $\text{rad } V = 0$, there exists an $A \in V$ such that $NA \neq 0$. The plane $U = \langle N, A\rangle$ is non-singular and (3.35) shows again $U^* = 0$, $V = U$.

THEOREM 3.7. *A space with orthogonal geometry is an orthogonal sum of lines*

$$V = \langle A_1\rangle \perp \langle A_2\rangle \perp \cdots \perp \langle A_n\rangle.$$

The A_i are called an orthogonal basis of V. V is non-singular if and only if none of the A_i are isotropic.

A non-singular symplectic space is an orthogonal sum of hyperbolic planes, in other words it is a hyperbolic space. Its dimension is always even.

THEOREM 3.8. *Let V be non-singular and U any subspace of V. Write $U = \text{rad } U \perp W$ and let $N_1, N_2, \cdots, N_r$ be a basis of* rad U. *Then we can find, in V, vectors $M_1, M_2, \cdots, M_r$ such that each N_i, M_i is a hyperbolic pair and such that the hyperbolic planes $P_i = \langle N_i, M_i \rangle$ are mutually orthogonal and also orthogonal to W. V will therefore contain the non-singular space*

$$\bar{U} = P_1 \perp P_2 \perp \cdots \perp P_r \perp W$$

which, in turn, contains U. Let finally σ be any isometry of U into some non-singular space V'. Then we can extend σ to an isometry $\bar{\sigma}$ of $\bar{U}$ into V'.

Proof:

1) For $r = 0$ nothing is to be proved. We may, therefore, use induction on r. The subspace

$$U_0 = \langle N_1, N_2, \cdots, N_{r-1} \rangle \perp W$$

is orthogonal to N_r but does not contain N_r. We have

$$\text{rad } U_0^* = \text{rad } U_0 = \langle N_1, N_2, \cdots, N_{r-1} \rangle.$$

This implies $N_r \in U_0^*$ but $N_r \notin \text{rad } U_0^*$. The space U_0^* must, therefore, contain a vector A such that $N_r A \neq 0$. The plane $\langle N_r, A \rangle$ is non-singular and contained in U_0^*. It is spanned by a hyperbolic pair N_r, M_r. Set $P_r = \langle N_r, M_r \rangle$. Since $P_r \subset U_0^*$, our space U_0 is orthogonal to P_r which shows that U_0 is contained in the non-singular space P_r^*. The radical of U_0 has only dimension $r - 1$. By induction hypothesis (used on P_r^* as the whole space) we can find hyperbolic pairs N_i, M_i in P_r^* (for $i \leq r - 1$) such that the P_i are mutually orthogonal and also orthogonal to W. Since they are orthogonal to P_r and since P_r is also orthogonal to W we have finished the construction of $\bar{U}$.

2) Let σ be the isometry of U into V', $N_i' = \sigma N_i$ and $W' = \sigma W$. Then $\sigma U = \langle N_1', N_2', \cdots, N_r' \rangle \perp W'$. We can, therefore, find in V' vectors M_i' such that N_i', M_i' is a hyperbolic pair and the planes $P_i' = \langle N_i', M_i' \rangle$ are mutually orthogonal and orthogonal to W'. The map σ of U into V' is now extended to a map $\bar{\sigma}$ of $\bar{U}$ into V' by prescribing $\bar{\sigma} M_i = M_i'$; $\bar{\sigma}$ is clearly an isometry.

THEOREM 3.9. (*Witt's theorem*). *Let V and V' be non-singular spaces which are isometric under some isometry ρ. Let σ be an isometry of a subspace U of V into V'. Then σ can be extended to an isometry of V onto V'.*

Proof:

1) If $\bar{U}$ is the non-singular subspace of V which was constructed in Theorem 3.8, then σ can be extended to an isometry of $\bar{U}$ into V'. This reduces the problem to the case where U is non-singular, an assumption which we shall make from now on.

2) For a symplectic space V let U' be the image of U under σ and write $V = U \perp U^*$ and $V' = U' \perp U'^*$. All we have really to do is to show that U^* and U'^* are isometric. But this is clear. V and V' have the same dimension since the isometry ρ exists. U^* and U'^* have, therefore, also the same dimension and they are non-singular; by Theorem 3.7 they are hyperbolic (we are in the symplectic case) and, hence, clearly isometric. We may assume from now on that our geometry is orthogonal.

3) Should our theorem be true for non-singular subspaces of lower dimension, then we can proceed as follows:

Assume that $U = U_1 \perp U_2$ where U_1 and U_2 are *proper* subspaces of U. We shall denote by U_1' the image of U_1 under σ and by τ the map $U_1 \to U_1'$ which σ induces by restriction. The map of U_1 into V' which σ induces can be extended to all of V by induction hypothesis; the space U_1^* will, thereby, be mapped onto the space $U_1'^*$. This means that U_1^* and $U_1'^*$ are isometric. Let us return to our σ. It will map the space U_2 (which is orthogonal to U_1) into the space $U_1'^*$; this mapping can (by induction hypothesis) be extended to an isometry λ of U_1^* onto $U_1'^*$ and now we are finished since $\tau \perp \lambda$ is the desired extension.

4) We have still to show that the theorem is true if U is non-singular and irreducible. $U = \langle A \rangle$ is then a line and $A^2 \neq 0$. For the image $C = \sigma A$ we have, therefore, $C^2 = A^2$. We must now remember that an isometry ρ of V onto V' was also given so that this vector C will be the image $\rho(B)$ of a vector B of V. We will know $A^2 = B^2$. If we can find an isometry τ of V onto V which carries A into B, then $\rho\tau$ will be an isometry of V onto V' which maps A onto C. To construct this τ we proceed as follows.

Since $(A + B)(A - B) = 0$ the vectors $A + B$ and $A - B$ are orthogonal. They can not both be isotropic since $2A = (A + B) +$

$(A - B)$ is *not* isotropic. Denote by $D = A + \epsilon B$ ($\epsilon = \pm 1$) a *non-isotropic* one of the vectors $A \pm B$. By what we have seen $A - \epsilon B$ belongs to the hyperplane $H = \langle A + \epsilon B \rangle^* = \langle D \rangle^*$. Let μ be the isometry which is identy on H and sends D into $-D$ (μ is an orthogonal sum of isometries on $\langle D \rangle$ and $\langle D \rangle^*$). Then

$$\mu(A + \epsilon B) = -A - \epsilon B,$$
$$\mu(A - \epsilon B) = A - \epsilon B.$$

Adding the two equations we get $\mu A = -\epsilon B$ (k is not of characteristic 2 in the orthogonal case).

If $\epsilon = -1$ we are done. If $\epsilon = +1$, then $\mu A = -B$. Let ν be the map which sends every vector X of V onto $-X$. It is obviously an isometry of V and $\nu\mu A = B$. This finishes the proof of our theorem.

Theorem 3.9 has many important consequences. An isotropic subspace U is naturally called maximal isotropic if U is not a proper subspace of some isotropic subspace of V.

THEOREM 3.10. *All maximal isotropic subspaces of a non-singular space V have the same dimension r. This invariant r of V is called the index of V.*

Proof: Let U_1 and U_2 be maximal isotropic and suppose $\dim U_1 \leq \dim U_2$. We can find a linear map of U_1 into U_2 which is an isomorphism into. Since both U_1 and U_2 are isotropic such a map is trivially an isometry. By Theorem 3.9 it can be extended to an isometry σ of V onto V. Thus $\sigma U_1 \subset U_2$ and consequently $U_1 \subset \sigma^{-1} U_2$. The space $\sigma^{-1} U_2$ is again isotropic. Since U_1 is maximal isotropic we must have $U_1 = \sigma^{-1} U_2$ which shows $\dim U_1 = \dim U_2$.

THEOREM 3.11. *The dimension of a maximal hyperbolic subspace is $2r$ where r is the index of V. Therefore $2r \leq n = \dim V$ and the maximal value $\frac{1}{2}n$ for the index r is reached if and only if V itself is hyperbolic (for instance in the symplectic case). A hyperbolic subspace H_{2s} is maximal hyperbolic if and only if the space H_{2s}^* does not contain any non-zero isotropic vectors. One achieves, therefore, a splitting $V = H_{2r} \perp W$ where W may be called anisotropic. The geometry of V determines uniquely the geometry of W (indepently of the choice of H_{2r}).*

Proof:

1) If U is a maximal isotropic subspace of V and $r = \dim U$, then Theorem 3.8 shows the existence of a hyperbolic space H_{2r} of dimen-

sion $2r$. Let $W = H_{2r}^*$ so that we have $V = H_{2r} \perp W$. Should W contain an isotropic vector $N \neq 0$, then N would be orthogonal to H_{2r}, hence to the subspace U, and not be contained in U. The space $U \perp \langle N \rangle$ would be isotropic, contradicting the fact that U was maximal isotropic.

2) If H'_{2s} is any hyperbolic subspace of V, then H'_{2s} has the form $\langle N'_1, M'_1 \rangle \perp \langle N'_2, M'_2 \rangle \perp \cdots \perp \langle N'_s, M'_s \rangle$ and $\langle N'_1, N'_2, \cdots, N'_s \rangle$ is an isotropic subspace of dimension s. This shows $s \leq r$. We can write $H_{2r} = H_{2s} \perp H_{2(r-s)}$ where H_{2s} and $H_{2(r-s)}$ are hyperbolic of dimensions $2s$ respectively $2(r - s)$. There exists an isometry σ of V onto V which carries H_{2s} onto H'_{2s}. Let $H'_{2(r-s)}$ and W' be the images of $H_{2(r-s)}$ respectively W under σ. Then

$$V = H'_{2s} \perp H'_{2(r-s)} \perp W',$$

and we see that $H'_{2(r-s)} \perp W'$ is the space orthogonal to H'_{2s}. If $s < r$, then this space will contain isotropic vectors; if $s = r$, then it reduces to W' which is isometric to W.

THEOREM 3.12. *Let V be non-singular. If U_1 and U_2 are isometric subspaces of V, then U_1^* and U_2^* are isometric.*

Proof: There is an isometry of V onto V which carries U_1 onto U_2. It carries, therefore, U_1^* onto U_2^*.

For several later applications we shall need a lemma about isometries of hyperbolic spaces.

THEOREM 3.13. *Let $V = \langle N_1, M_1 \rangle \perp \langle N_2, M_2 \rangle \perp \cdots \perp \langle N_r, M_r \rangle$ be a hyperbolic space and σ an isometry of V onto V which keeps each vector N_i fixed. Then σ is a rotation. As a matter of fact σ is of the following form:*

$$\sigma N_i = N_i, \qquad \sigma M_j = \sum_{\nu=1}^{r} N_\nu a_{\nu j} + M_j$$

where the matrix (a_{ij}) is symmetric if the geometry is symplectic and skew symmetric if it is orthogonal.

Proof: We have of course $\sigma N_i = N_i$. Let

$$\sigma M_j = \sum_{\nu=1}^{r} N_\nu a_{\nu j} + \sum_{\mu=1}^{r} M_\mu b_{\mu j} .$$

We find

$$\sigma N_i \cdot \sigma M_j = N_i \cdot \sigma M_j = b_{ij}$$

and, since we must have $\sigma N_i \cdot \sigma M_j = N_i M_j$ if σ is an isometry, we get $b_{ii} = 1$ and $b_{ij} = 0$ for $i \neq j$. Thus

$$\sigma M_j = \sum_{\nu=1}^{r} N_\nu a_{\nu j} + M_j \ .$$

We must still check that $\sigma M_i \cdot \sigma M_j = M_i M_j = 0$. This leads to

$$\left(\sum_{\nu=1}^{r} N_\nu a_{\nu i} + M_i \right)\left(\sum_{\mu=1}^{r} N_\mu a_{\mu j} + M_j \right) = a_{ji} + (M_i N_i) a_{ij} = 0.$$

If V is symplectic, then $M_i N_i = -1$ and we get $a_{ji} = a_{ij}$. If V is orthogonal, then $a_{ji} = -a_{ij}$ and since we assume in this case that the characteristic of k is $\neq 2$ we see that (a_{ij}) is skew symmetric.

Obviously, in both cases, $\det \sigma = +1$.

We shall now turn to isometries of general non-singular spaces. The identity shall be denoted by 1 or by 1_V if a reference to the space is needed. The map σ which sends each vector X onto $-X$ satisfies $(1 + \sigma)X = X - X = 0$, $1 + \sigma = 0$ and shall, therefore, be denoted by -1 respectively -1_V if a reference to the space V is needed.

Let $V = U \perp W$. Then we can form the isometry $\sigma = -1_U \perp 1_W$. It is obviously of interest only if the characteristic of k is $\neq 2$ since otherwise it would be merely 1_V . If $X = A + B$ with $A \in U$ and $B \in W$, then $\sigma X = -A + B$. We have $\sigma X = X$ if and only if $A = 0$, or $X \in W$, and $\sigma X = -X$ if and only if $B = 0$ or $X \in U$. This means that U and W are characterised by σ. Such a map σ satisfies obviously $\sigma^2 = 1$. We shall now prove

THEOREM 3.14. *An isometry σ of a non-singular space V is called an involution if $\sigma^2 = 1$. If the characteristic of k is $\neq 2$ then every involution is of the form $-1_U \perp 1_W$ (resulting from a splitting $V = U \perp W$). If the characteristic of k is 2, each involution of V will keep every vector of a certain maximal isotropic subspace of V fixed and be of the type of isometries discussed in Theorem 3.13.*

Proof:

1) Suppose $\sigma^2 = 1$. Then $XY = \sigma X \cdot \sigma Y$ and $\sigma X \cdot Y = \sigma^2 X \cdot \sigma Y = X \cdot \sigma Y$. Consequently

$$(\sigma X - X)(\sigma Y + Y) = \sigma X \cdot \sigma Y + \sigma X \cdot Y - X \cdot \sigma Y - XY = 0.$$

The two subspaces $U = (\sigma - 1)V$ and $W = (\sigma + 1)V$ are, therefore, orthogonal.

2) Let the characteristic of k be $\neq 2$. A vector $\sigma X - X$ of U is reversed by σ, a vector $\sigma X + X$ of W is preserved by σ. Consequently $U \cap W = 0$. Since any vector X can be written as $-\frac{1}{2}(\sigma X - X) + \frac{1}{2}(\sigma X + X)$ we see that $V = U \perp W$ and $\sigma = -1_U \perp 1_W$.

3) If the characteristic of k is 2 we are in the symplectic case. The two maps $\sigma - 1$ and $\sigma + 1$ are the same and 1) means now that $(\sigma + 1)V$ is isotropic. Let K be the kernel of the map $\sigma - 1 = \sigma + 1$, hence the subspace of all vectors of V which are left fixed by σ. Then $(\sigma + 1)\, V \subset K$. Let us write

$$K = \text{rad } K \perp K_0 .$$

Since $\sigma K_0 = K_0$, we have $\sigma K_0^* = K_0^*$ and thus $(\sigma - 1)\, K_0^* \subset K_0^*$. The only vectors of K which are contained in K_0^* are those of rad K which means that rad K is the kernel of $\sigma - 1$ if applied to K_0^*. This shows

$$\dim K_0^* = \dim\, (\sigma - 1)K_0^* + \dim \text{rad } K.$$

Both $(\sigma - 1)K_0^*$ and rad K are isotropic subspaces of the non-singular space K_0^* and have, therefore, dimensions at most $\frac{1}{2} \dim K_0^*$. We see that this value is reached, i.e., that rad K must be a maximal isotropic subspace of K_0^*. If K_1 denotes a maximal isotropic subspace of K_0, then $K_2 = \text{rad } K \perp K_1$ is a maximal isotropic subspace of V. But $K_2 \subset K$ shows that every vector of K_2 is left fixed by σ.

DEFINITION 3.10. Suppose the characteristic of k is $\neq 2$. If $\sigma = -1_U \perp 1_W$ and $p = \dim U$, then we call p the type of the involution σ. We have obviously $\det \sigma = (-1)^p$. Since U has to be non-singular the type p must be an even number if V is symplectic; if V has orthogonal geometry, then p might be any number $\leq n = \dim V$. An involution of type $p = 1$ shall be called a symmetry with respect to the hyperplane W. An involution of type 2 shall be called a 180° rotation.

The reason for the word symmetry should be clear: $U = \langle A \rangle$ is a non-singular line, $U^* = W$ a non-singular hyperplane and the image of a vector $xA + B$ (with $B \varepsilon W$) is $-xA + B$.

The following theorem characterizes the isometries $\pm 1_V$.

THEOREM 3.15. *Let V be non-singular and σ an isometry of V which keeps all lines of V fixed. Then $\sigma = \pm 1_V$.*

Proof: If σ keeps the line $\langle X \rangle$ fixed, then $\sigma X = Xa$ and for any

$Y \in \langle X \rangle$ we have $\sigma Y = \sigma(Xb) = \sigma(X) \cdot b = Xab = Ya$. This a may still depend on the line $\langle X \rangle$ if σ keeps every line of V fixed. If $\langle X \rangle$ and $\langle Y \rangle$ are different lines, then X and Y are independent vectors. We have on one hand $\sigma(X + Y) = (X + Y) \cdot c$ and on the other $\sigma(X + Y) = \sigma(X) + \sigma(Y) = Xa + Yb$. A comparison shows $a = c = b$ and we know now that we have $\sigma X = Xa$ with the same a for all X. Let X and Y be vectors such that $XY \neq 0$. Then $XY = \sigma X \cdot \sigma Y = Xa \cdot Ya = (XY)a^2$. We see $a^2 = 1$, $a = \pm 1$.

4. *Special features of orthogonal geometry*

In this section V will stand for a *non-singular* space of dimension n with an orthogonal geometry.

DEFINITION 3.11. The group of all isometries of V into V is called the orthogonal group of V and is denoted by O. The subgroup of all rotations is denoted by O^+ and the commutator subgroup of O by Ω. If it is necessary to be more specific in these symbols, then we shall put n as a subscript and inclose in a parenthesis any further information. For instance $O_n(k, f)$ denotes the orthogonal group of a space over k whose dimension is n and whose geometry is based on the quadratic form f. O^+ is the kernel of the determinant map; the image is commutative, hence $\Omega \subset O^+$. The index of O^+ in O is 2 since reflexions (for instance symmetries) are present if $n \geq 1$.

We state a slight improvement of Witt's theorem.

THEOREM 3.16. *Let σ be an isometry of a subspace U of V into V. It is possible to prescribe the value ± 1 for the determinant of an extension τ of σ to all of V if and only if* $\dim U + \dim \operatorname{rad} U < n$.

Proof:

1) If τ_1 and τ_2 are two extensions of σ to V whose determinants have opposite sign, then $\rho = \tau_1^{-1}\tau_2$ will be a reflexion of V which keeps every vector of U fixed. Conversely, if ρ is a reflexion which keeps every vector of U fixed and τ_1 one extension of σ to V, then $\tau_2 = \tau_1\rho$ is another extension and $\det \tau_2 = -\det \tau_1$. All we have to do is to see whether such a reflexion exists.

2) Suppose $\dim U + \dim \operatorname{rad} U < n$. The space $\bar{U}$ of Theorem 3.8 is of dimension $\dim U + \dim \operatorname{rad} U$. We can write $V = \bar{U} \perp \bar{U}^*$ where $\bar{U}^* \neq 0$. Let λ be a reflexion of $\bar{U}^*$ and put $\rho = 1_{\bar{U}} \perp \lambda$.

3) Suppose $\dim U + \dim \operatorname{rad} U = n$. Then $V = \bar{U} = H_{2r} \perp W$.

We are looking for a ρ which is identity on U and, therefore, on W. It must map $W^* = H_{2r}$ onto itself and be a reflexion on H_{2r}. But it should keep every vector of rad U (which is a maximal isotropic subspace of H_{2r}) fixed. Theorem 3.13 shows that such a ρ does not exist.

THEOREM 3.17. *Let σ be an isometry of V which leaves every vector of a hyperplane H fixed. If H is singular, then $\sigma = 1$. If H is non-singular, then σ is either identity or the symmetry with respect to H. The effect of an isometry on a hyperplane H determines the isometry completely if H is singular, and up to a symmetry with respect to H if H is non-singular.*

Proof:

1) Let H be singular. The line $L = H^* = \langle N \rangle$ is also singular, N is isotropic. We have rad H = rad $H^* = \langle N \rangle$ and can write

$$H = \langle N \rangle \perp W$$

where W is non-singular and of dimension $n - 2$. The plane W^* is non-singular and contains N. There exists, therefore, a *unique* hyperbolic pair N, M in W^*. If σ leaves H element-wise fixed then $\sigma W^* = W^*$. The pair $\sigma N, \sigma M$ is again hyperbolic and $\sigma N = N$; the uniqueness gives now $\sigma M = M$ which shows σ to be identity on W^* as well as on W.

2) Let H be non-singular. The line $L = H^* = \langle A \rangle$ will be non-singular and $V = L \perp H$. If σ is identity on H then $\sigma = \lambda \perp 1_H$ and λ, as isometry on L, must be $\pm 1_L$.

3) If σ and τ have the same effect on H, then $\sigma^{-1}\tau$ leaves every vector of H fixed. If H is singular, then $\sigma^{-1}\tau = 1$, $\sigma = \tau$. If H is non-singular, $\sigma^{-1}\tau$ may be the symmetry with respect to H.

We can improve on Theorem 3.15 in two ways.

THEOREM 3.18. *Suppose that $n \geq 3$ and that V contains isotropic lines. If σ keeps every isotropic line fixed, then $\sigma = \pm 1_V$.*

Proof: V contains a hyperbolic pair N, M. The space $\langle N, M \rangle^*$ is non-singular and $\neq 0$. Let $B \in \langle N, M \rangle^*$. If $B^2 = 0$ then $\sigma B = Bc$ by assumption. Assume now $B^2 \neq 0$. Then $\langle N, M \rangle \perp \langle B \rangle = \langle N, M, B \rangle$ is non-singular and $\langle B \rangle$ the only line in $\langle N, M, B \rangle$ which is orthogonal to $\langle N, M \rangle$. If we can prove that σ maps $\langle N, M, B \rangle$ onto itself, then we obtain again $\sigma B = Bc$ since, by assumption, $\sigma N = Na$ and $\sigma M = Mb$.

Regardless of whether $B^2 = 0$ or not, the vector $N - \frac{1}{2}B^2 \cdot M + B$ is easily seen to be isotropic. By assumption we have, therefore, on one hand

$$\sigma\left(N - \frac{1}{2}B^2 \cdot M + B\right) = d\left(N - \frac{1}{2}B^2 \cdot M + B\right)$$

$$= dN - \frac{d}{2}B^2 \cdot M + dB$$

and on the other hand

$$\sigma\left(N - \frac{1}{2}B^2 \cdot M + B\right) = aN - \frac{b}{2}B^2 \cdot M + \sigma(B).$$

A comparison shows first that $\sigma(B) \in \langle N, M, B\rangle$ which proves $\sigma(B) = Bc$ also if $B^2 \neq 0$, and secondly that $a = d = c$. If B is selected so that $B^2 \neq 0$, then one sees in addition that $a = d = b$. This proves now that every vector of $\langle N, M\rangle$ and every vector of $\langle N, M\rangle^*$ is multiplied by a. Therefore $\sigma X = Xa$ for all $X \in V$ and Theorem 3.15 shows $\sigma = \pm 1_V$.

THEOREM 3.19. *Suppose that σ leaves every non-isotropic line of V fixed but omit the single case where V is a hyperbolic plane over a field with 3 elements. Then $\sigma = \pm 1_V$.*

Proof: Because of Theorem 3.15 we may assume that V contains a hyperbolic pair N, M.

1) $\dim V \geq 3$; then $\dim\langle N, M\rangle^* \geq 1$. Let B be a non-isotropic vector of $\langle N, M\rangle^*$. Both B and $N + B$ are non-isotropic, hence, by assumption, $\sigma B = Bb$ and $\sigma(N + B) = (N + B)c = \sigma(N) + Bb$; therefore $\sigma(N) = Nc + B(c - b)$. But $\sigma(N)$ must be isotropic; we deduce $c = b$, $\sigma(N) = Nc$ and σ keeps also the isotropic lines fixed. We use now Theorem 3.15.

2) $V = \langle N, M\rangle$. If the lines $\langle N\rangle$, $\langle M\rangle$ are not left fixed, then σ must interchange them since they are the only isotropic lines of V. Hence $\sigma N = aM$, $\sigma M = bN$. From $NM = \sigma N \cdot \sigma M = aM \cdot bN = 1$ we deduce $b = 1/a$. Let c be distinct from $0, \pm a$. The vector $N + cM$ is not isotropic, hence, by assumption,

$$\sigma(N + cM) = d(N + cM) = dN + dcM$$

on one hand and

$$\sigma(N + cM) = aM + \frac{c}{a}N$$

on the other. A comparison yields $a = dc$, $c/a = d$, hence $c^2 = a^2$, a contradiction.

In our next theorem, due to E. Cartan and J. Dieudonné, we try to express each isometry of V by symmetries with respect to hyperplanes.

THEOREM 3.20. *Let V be non-singular and* $\dim V = n$. *Every isometry of V onto V is a product of at most n symmetries with respect to non-singular hyperplanes.*

Proof: The theorem is trivial for $\sigma = 1$ and also for $n = 1$. We use induction on n and have to distinguish 4 cases.

Case 1). There exists a non-isotropic vector A, left fixed by σ. Let $H = \langle A \rangle^*$; then $\sigma H = H$. Let λ be the restriction of σ to H and write $\lambda = \tau_1 \tau_2 \cdots \tau_r$ with $r \leq n - 1$ (by induction hypothesis) where τ_i is a symmetry of the space H with respect to a hyperplane H_i of H. Put $\bar{\tau}_i = 1_L \perp \tau_i$ where $L = \langle A \rangle$. Each $\bar{\tau}_i$ leaves the hyperplane $L \perp H_i$ of V fixed and is, therefore (as reflexion), a symmetry of V. The product $\bar{\tau}_1 \bar{\tau}_2 \cdots \bar{\tau}_r$ is $1_L \perp \lambda = \sigma$ and we know in this case that σ can be expressed by at most $n - 1$ symmetries.

Case 2). There exists a non-isotropic vector A such that $\sigma A - A$ is *not* isotropic. Let $H = \langle \sigma A - A \rangle^*$ and let τ be the symmetry with respect to H. Since $(\sigma A + A)(\sigma A - A) = (\sigma A)^2 - A^2 = 0$ (σ is an isometry) we have $\sigma A + A \in H$. Therefore

$$\tau(\sigma A + A) = \sigma A + A, \qquad \tau(\sigma A - A) = A - \sigma A.$$

Adding we have: $\tau\sigma(2A) = 2A$ which shows that $\tau\sigma$ leaves A fixed. By case 1) $\tau\sigma = \tau_1 \tau_2 \cdots \tau_r$ where $r \leq n - 1$. Since $\tau^2 = 1$, we obtain, after multiplication by τ from the left, $\sigma = \tau\tau_1\tau_2 \cdots \tau_r$, a product of at most n symmetries.

Case 3). $n = 2$. Cases 1) and 2) allow us to assume that V contains non-zero isotropic vectors: $V = \langle N, M \rangle$ where N, M is a hyperbolic pair.

a) $\sigma N = aM$, $\sigma M = a^{-1}N$. Then $\sigma(N + aM) = aM + N$ is a fixed non-isotropic vector. We are in Case 1).

b) $\sigma N = aN$, $\sigma M = a^{-1}M$. We may assume $a \neq 1$ since $a = 1$ means $\sigma = 1$. $A = N + M$ and $\sigma A - A = (a - 1)N + (a^{-1} - 1)M$ are not isotropic which brings us back to Case 2).

Case 4). We can now assume that $n \geq 3$, that no non-isotropic vector is left fixed, and, finally, that $\sigma A - A$ is isotropic whenever A is not isotropic.

Let N be an isotropic vector. The space $\langle N \rangle^*$ has at least dimension 2 and its radical is the same as that of $\langle N \rangle$ (i.e., it is $\langle N \rangle$). $\langle N \rangle^*$ contains a non-isotropic vector A.

We have $A^2 \neq 0$ and $(A + \epsilon N)^2 = A^2 \neq 0$. We conclude by our assumption that $\sigma A - A$ as well as the vectors

$$\sigma(A + \epsilon N) - (A + \epsilon N) = (\sigma A - A) + \epsilon(\sigma N - N)$$

are isotropic. The square is, therefore,

$$2\epsilon(\sigma A - A)(\sigma N - N) + \epsilon^2(\sigma N - N)^2 = 0.$$

This last equation is written down for $\epsilon = 1$ and $\epsilon = -1$ and added. We get $2(\sigma N - N)^2 = 0$ or $(\sigma N - N)^2 = 0$.

We know, therefore, that $\sigma X - X$ will be isotropic whether X is isotropic or not. The set W of all these vectors $\sigma X - X$ is the image of V under the map $\sigma - 1$. It contains only isotropic vectors and is, consequently an isotropic subspace of V. A product of any two of its vectors will be zero.

Let now $X \in V$ and $Y \in W^*$. Consider

$$(\sigma X - X)(\sigma Y - Y) = 0 = \sigma X \cdot \sigma Y - X \cdot \sigma Y - (\sigma X - X) \cdot Y.$$

But $\sigma X - X \in W$, $Y \in W^*$ so that the last term is 0. Furthermore, $\sigma X \cdot \sigma Y = XY$ since σ is an isometry. Hence

$$X(Y - \sigma Y) = 0$$

which is true for all $X \in V$. This means $Y - \sigma Y \in \operatorname{rad} V = 0$ or $Y = \sigma Y$.

We see that every vector of W^* is left fixed. We had assumed that a non-isotropic vector is not left fixed. This shows that W^* is an isotropic subspace. We know $\dim W \leq \frac{1}{2}n$ and $\dim W^* \leq \frac{1}{2}n$ since these spaces are isotropic. From $\dim W + \dim W^* = n$ we conclude that the equality sign holds. The space V is, therefore, a hyperbolic space H_{2r}, $n = 2r$ and W^* a maximal isotropic subspace of H_{2r}. The isometry σ leaves every vector of W^* fixed, and Theorem 3.13 shows σ to be a *rotation*.

We can conclude, for instance, that for the space $V = H_{2r}$ our theorem holds at least for reflexions. Let now τ be any symmetry of $V = H_{2r}$. Then $\tau\sigma$ is a *reflexion* of H_{2r}, hence $\tau\sigma = \tau_1\tau_2 \cdots \tau_s$ with $s \leq n = 2r$, but since $\tau\sigma$ is a reflexion s must be odd, hence $s \leq 2r - 1$. We get $\sigma = \tau\tau_1\tau_2 \cdots \tau_s$, a product of $s + 1 \leq 2r = n$ symmetries.

REMARK 1. Let $\sigma = \tau_1\tau_2 \cdots \tau_s$ where τ_i is the symmetry with respect to the hyperplane H_i. Let $U_i = H_1 \cap H_2 \cap \cdots \cap H_i$. Then $\operatorname{codim} U_i + \operatorname{codim} H_{i+1} = \operatorname{codim} U_{i+1} + \operatorname{codim}(U_i + H_{i+1})$. Hence

$$\operatorname{codim} U_{i+1} \leq \operatorname{codim} U_i + 1.$$

This shows $\operatorname{codim} U_i \leq i$, especially $\operatorname{codim}(H_1 \cap H_2 \cap \cdots \cap H_s) \leq s$ or

$$\dim(H_1 \cap H_2 \cap \cdots \cap H_s) \geq n - s.$$

Every vector of $H_1 \cap H_2 \cap \cdots \cap H_s$ is left fixed by every τ_i and, therefore, by σ. This shows that every vector of a subspace of a dimension at least $n - s$ is left fixed by σ. Should $s < n$, then σ leaves some non-zero vector fixed. We conclude that an isometry without a fixed non-zero vector (as for instance -1_V) can not be written as product of less than n symmetries.

If σ is a product of n symmetries, then $\det \sigma = (-1)^n$. Thus we state:

If n is even, then every reflexion keeps some non-zero vector fixed, and if n is odd, the same holds for a rotation.

REMARK 2. Suppose that σ can not be written by less than n symmetries, and let τ be an arbitrary symmetry. Then $\det(\tau\sigma) = -\det \sigma$, which means $\tau\sigma = \tau_1\tau_2 \cdots \tau_s$ with $s < n$ and, hence, $\sigma = \tau\tau_1 \cdots \tau_s$ $(s + 1 \leq n)$. We see that the first factor in the product for σ can be an arbitrarily chosen symmetry.

Examples

I. $\dim V = 2$.

Every reflexion is a symmetry and every rotation is a product of two symmetries in which the first factor can be chosen in advance (also for $1 = \tau\tau$). Theorem 3.17 shows that a rotation $\sigma \neq 1$ can leave only the zero vector fixed and that any rotation is uniquely determined by the image of one non-zero vector.

Let τ be a symmetry, $\sigma = \tau\tau_1$ a rotation. Then $\tau\sigma\tau^{-1} = \tau^2\tau_1\tau = \tau_1\tau = (\tau\tau_1)^{-1} = \sigma^{-1}$. We have, therefore

$$\tau\sigma\tau^{-1} = \sigma^{-1}. \tag{3.36}$$

If $\sigma_1 = \tau\tau_2$ is another rotation, we have

$$\sigma_1\sigma\sigma_1^{-1} = \tau(\tau_2\sigma\tau_2^{-1})\tau^{-1} = \tau\sigma^{-1}\tau^{-1} = \sigma.$$

Thus the group O_2^+ is commutative and the interaction with the coset of symmetries is given by (3.36).

Theorem 3.14 tells us that $\pm 1_V$ are the only involutions among the rotations, if $n = 2$.

This also allows us to answer the question when the group O_2 itself is abelian. Should this be the case, then (3.36) shows $\sigma = \sigma^{-1}$ or $\sigma^2 = 1$, i.e., that there are only two rotations. O_2^+ has index 2 in O_2, τO_2^+ is the other coset (τ a symmetry) so that there are only two symmetries in such a geometry. This means that V contains only two non-isotropic lines. If A, B is an orthogonal basis, then $\langle A \rangle$ and $\langle B \rangle$ are the only non-isotropic lines. Hence the two lines $\langle A + B \rangle$ and $\langle A - B \rangle$ are isotropic and, since V can only contain two isotropic lines, we have enumerated all lines of V. The field k has only three elements. Conversely, if V is a hyperbolic plane over a field with three elements, then V contains four lines, two isotropic and two non-isotropic ones. There are two symmetries and thus two rotations. These rotations can only be $\pm 1_V$ and we have $\sigma = \sigma^{-1}$ for every rotation; O_2 is commutative. To summarize:

If $\dim V = 2$, but if V is not a hyperbolic plane over a field with three elements, then O_2 is not commutative and O_2^+ contains rotations σ for which $\sigma^2 \neq 1$.

II. $\dim V = 3$.

Theorem 3.17 shows that a rotation $\neq 1$ can not leave two independent vectors fixed. However it is a product of two symmetries and leaves, therefore, all vectors of some line $\langle A \rangle$ fixed. We call $\langle A \rangle$ the axis of the rotation.

Let σ be an isometry which keeps one non-zero vector A fixed and reverses some other non-zero vector B. The vectors A, B are independent. σ^2 will be a rotation which keeps every vector of $\langle A, B \rangle$ fixed. By Theorem 3.17 we have $\sigma^2 = 1$ (but $\sigma \neq \pm 1$). It follows that σ is an involution $\neq \pm 1$, hence a symmetry, if it is a reflexion, and a 180° rotation, if it is a rotation.

If a rotation σ is written as product $\tau_1 \tau_2$ of symmetries, we can also write $\sigma = (-\tau_1)(-\tau_2)$ as product of two 180° rotations.

Let us look finally at all rotations with given axis $\langle A \rangle$ (include also $\sigma = 1$). They form obviously a subgroup of O_3^+. We have to distinguish two cases.

1) $\langle A \rangle$ non-isotropic. We can write $V = \langle A \rangle \perp \langle A \rangle^*$ and our group is obviously isomorphic to the group O_2^+ of $\langle A \rangle^*$.

2) $A = N$, $N^2 = 0$. We have rad $\langle N\rangle^* =$ rad $\langle N\rangle = \langle N\rangle$. Hence $\langle N\rangle^* = \langle N, B\rangle$ with $NB = 0$, $B^2 \neq 0$. We shall determine all isometries σ of V for which $\sigma N = N$. Then $\sigma\langle N\rangle^* = \langle N\rangle^*$ and σ is described completely by the effect it has on the singular plane $\langle N\rangle^*$ (Theorem 3.17). We set $\sigma B = xN + yB$ and from $B^2 = (\sigma B)^2 = y^2B^2$ get $y = \pm 1$.

Let τ_x be the symmetry with respect to the plane perpendicular to $\frac{1}{2}xN - B$. Since $N\cdot(\frac{1}{2}xN - B) = 0$ we have $\tau_x N = N$; $\tau_x B = \tau_x(\frac{1}{2}xN - (\frac{1}{2}xN - B)) = \frac{1}{2}xN + \frac{1}{2}xN - B = xN - B$. For $x = 0$ we have $\tau_0 N = N$, $\tau_0 B = -B$. If we denote $\tau_0\tau_x$ by σ_x, then σ_x is a rotation and

$$\sigma_x N = N, \qquad \sigma_x B = xN + B.$$

The symmetries τ_x and the rotations σ_x are now all isometries which keep N fixed. Obviously

$$\sigma_x\sigma_y = \sigma_{x+y}$$

which shows that the rotations with axis $\langle N\rangle$ form a group isomorphic to the additive group of k.

We remark that $\sigma_x = (\sigma_{x/2})^2$ which shows that every σ_x is the square of a rotation.

Let us look also at reflexions. If σ is a reflexion, then $-\sigma$ is a rotation. Should $\sigma \neq -1$, then $-\sigma$ will have an axis $\langle A\rangle$. This means that σ reverses the vectors of precisely one line $\langle A\rangle$. Should σ keep some non-zero vector fixed, then σ is a symmetry, as we have seen earlier. The reflexions without fixed vectors are exactly those that can not be written as product of less than three symmetries.

Some of these results are useful if $n \geq 3$.

THEOREM 3.21. *Let* dim $V \geq 3$. *The centralizer of the set of squares of rotations consists of* $\pm 1_V$.

Proof:

1) Suppose V contains an isotropic line $\langle N\rangle$. Then rad $\langle N\rangle^* =$ rad $\langle N\rangle = \langle N\rangle$ and we can write

$$\langle N\rangle^* = \langle N\rangle \perp W$$

where W is non-singular and of dimension $n - 2 \geq 1$. Let U be a non-singular subspace of W of dimension $n - 3$ (possibly $U = 0$) and write $V = U^* \perp U$. This space U^* is non-singular, of dimension 3, and contains $\langle N\rangle$. Let σ_x be a rotation of U^* with axis $\langle N\rangle$; σ_x is the square of the rotation $\sigma_{x/2}$. If we put $\sigma = \sigma_{x/2} \perp 1_U$, then $\sigma^2 = \sigma_x \perp 1_U$. Which vectors of V are left fixed by σ^2? Let $A \in U^*$, $B \in U$ and suppose $A + B$ is left fixed by $\sigma_x \perp 1_U$. Obviously A must be left fixed by σ_x and this means $A \in \langle N \rangle$. The fixed vectors are, therefore, those of $\langle N \rangle \perp U = W_0$, a subspace whose radical is $\langle N \rangle$. Let now τ be commutative with all squares of rotations. Then $\tau\sigma^2\tau^{-1} = \sigma^2$; $\tau\sigma^2\tau^{-1}$ leaves every vector of τW_0 fixed so that $\tau W_0 = W_0$. This means that the radical $\tau\langle N \rangle$ of τW_0 is $\langle N \rangle$. Our τ leaves every isotropic line $\langle N \rangle$ fixed and Theorem 3.19 shows $\tau = \pm 1_V$.

2) Suppose V does not contain isotropic vectors. Let P be any plane of V. We know that there exists a rotation λ of P such that $\lambda^2 \neq 1$. Write $V = P \perp P^*$ and let $\sigma = \lambda \perp 1_{P^*}$. Then σ^2 leaves only the vectors of P^* fixed. If $\tau\sigma^2\tau^{-1} = \sigma^2$, then we see as before that $\tau P^* = P^*$ and consequently $\tau P = P$. The element τ leaves every plane fixed. A given line is an intersection of two planes (since $n \geq 3$) and we see that τ leaves every line of V fixed. From Theorem 3.15 we see now $\tau = \pm 1_V$.

The meaning of Theorem 3.21 will become clear a little later.

Let now τ_1 and τ_2 be symmetries with respect to the non-singular hyperplanes H_1 and H_2. Assume again $n \geq 3$. Let $\langle A \rangle = H_1^*$ and $\langle B \rangle = H_2^*$. $\langle A \rangle$ and $\langle B \rangle$ are non-singular, the space $P = \langle A, B \rangle$ (of dimension ≤ 2) not isotropic, its radical of dimension ≤ 1. We have again rad $P^* =$ rad P, P^* has dimension at least $n - 2$ and contains a non-singular subspace U of dimension $n - 3$. The space $U \subset H_1 \cap H_2$, since U is orthogonal to both A and B. τ_1 and τ_2 are identity on U and symmetries $\bar{\tau}_1$, $\bar{\tau}_2$ on the three dimensional non-singular space U^*. We can write

$$\tau_1\tau_2 = (-\bar{\tau}_1 \perp 1_U)(-\bar{\tau}_2 \perp 1_U)$$

and have replaced $\tau_1\tau_2$ by a product of two 180° rotations.

Let now $\sigma = \tau_1\tau_2 \cdots \tau_s$ be any rotation of V, expressed as product of $s \leq n$ symmetries. Since s must be even, we can group the product in pairs and replace them by 180° rotations. Thus we have

THEOREM 3.22. *If $n \geq 3$, then every rotation is a product of an even number $\leq n$ of 180° rotations.*

For our next aim we need a group theoretical lemma.

LEMMA. *Let G be a group and S a subset of elements with the following properties:*

1) *G is generated by S.*
2) *For $x \varepsilon G$ and $s \varepsilon S$ we have $xsx^{-1} \varepsilon S$.*
3) *$s^2 = 1$ for all $s \varepsilon S$.*

Then the commutator subgroup G' of G is generated by the commutators $s_1 s_2 s_1^{-1} s_2^{-1} = (s_1 s_2)^2$, $s_1 \varepsilon S$, $s_2 \varepsilon S$. G' contains the squares of all elements of G.

Proof: Let H be the set of all products of elements of the form $s_1 s_2 s_1^{-1} s_2^{-1}$ with $s_1, s_2 \varepsilon S$. H is obviously a group and property 2) shows that it is invariant in G.

Let f be the canonical map $G \to G/H$ with kernel H. Since $s_1 s_2 s_1^{-1} s_2^{-1} \varepsilon H$ we have $f(s_1 s_2 s_1^{-1} s_2^{-1}) = 1$, hence $f(s_1)f(s_2) = f(s_2)f(s_1)$. The set S generates G so that the image $f(G) = G/H$ is generated by $f(S)$. We see that $f(G)$ is a commutative group which shows that G' is contained in the kernel H. Trivially, since H is generated by commutators, $H \subset G'$. Every generator of $f(G)$ has at most order 2 so that every element in $f(G)$ has at most order 2. This proves $f(x^2) = 1$, $x^2 \varepsilon H$.

If σ is a symmetry (180° rotation) with respect to the hyperplane H (with the plane P) then $\tau\sigma\tau^{-1}$ is again a symmetry (180° rotation) with hyperplane τH (with plane τP) and $\sigma^2 = 1$. The set of all symmetries (180° rotations) generates the group O (the group O^+) and satisfies the conditions of the lemma.

For $n \geq 3$ the commutator group of O_n^+ is generated by all $(\sigma_1\sigma_2)^2$, where σ_1 and σ_2 are 180° rotations, and it contains all squares of rotations. The commutator group Ω_n of O_n is generated by all $(\tau_1\tau_2)^2$ where τ_1 and τ_2 are symmetries; since $\tau_1\tau_2$ is a rotation, $(\tau_1\tau_2)^2$ is a square of a rotation. This shows that Ω_n is not larger than the commutator group of O_n^+. The meaning of Theorem 3.21 becomes clear. We have determined the centralizer of the group Ω_n, therefore, also the centralizer of O_n^+ and O_n (since $\pm 1_V$ obviously commute with every element of O_n). Thus we have

THEOREM 3.23. *Let $n \geq 3$. The groups O_n and O_n^+ have the same commutator group Ω_n which is generated by all $(\tau_1\tau_2)^2$ where τ_1 and τ_2 are symmetries and contains all squares of elements of O_n. The centralizer of O_n, O_n^+ or Ω_n is $\pm 1_V$. The center of O_n is $\pm 1_V$. If n is odd, then 1_V is the center of O_n^+ and of Ω_n (-1_V is a reflexion). If n is even, then $\pm 1_V$ is the center of O_n^+; the center of Ω_n is $\pm 1_V$ or 1_V depending on whether $-1_V \varepsilon \Omega_n$ or not. In the factor groups O_n/Ω_n and O_n^+/Ω_n all elements have at most order 2.*

Another important property of the group Ω_n is expressed in

THEOREM 3.24. *If $n \geq 3$, then Ω_n is irreducible. This means that no proper subspace $U \neq 0$ is mapped into itself by all elements of Ω_n.*

Proof: Suppose that there is such a proper subspace $U \neq 0$. We shall derive a contradiction and distinguish two cases.

1) U is non-singular. Write $V = U \perp W$; if $\sigma \varepsilon \Omega_n$, then, by assumption, $\sigma U = U$ and hence $\sigma = \tau \perp \rho$ with $\tau \varepsilon O(U)$, $\rho \varepsilon O(W)$. The involution $\lambda = -1_U \perp 1_W$ would satisfy $\sigma\lambda = \lambda\sigma$ and this implies that λ is in the centralizer of Ω_n, which is a contradiction to Theorem 3.23.

2) U is singular. Then rad $U \neq 0$ is also mapped into itself by Ω_n and, replacing U by rad U, we may assume that U is isotropic. Let $N \neq 0$ be a vector of U and N, M a hyperbolic pair. Since $n \geq 3$ we can find a three-dimensional non-singular subspace $W = \langle N, M, A \rangle$ where A is a non-isotropic vector of $\langle N, M \rangle^*$. Isotropic subspaces of W have at most dimension 1 so that $U \cap W = \langle N \rangle$ (remember that U is isotropic). Let $\sigma_x \neq 1$ be a rotation of W with axis $\langle M \rangle$. If σ_x would keep the line $\langle N \rangle$ fixed, then it would induce a rotation of the plane $\langle N, M \rangle$; but $\sigma_x M = M$ would then imply $\sigma_x N = N$ and consequently $\sigma_x = 1_W$ (two vectors are kept fixed). The line $\langle N \rangle$ is, therefore, moved into another line of W, hence into a line *not* contained in U. Setting $\sigma = \sigma_x \perp 1_{W^*}$ we have $\sigma \varepsilon \Omega_n$ since $\sigma_x = (\sigma_{x/2})^2$, and $\sigma N \not\varepsilon U$.

EXERCISES.

1) Investigate the irreducibility of the groups O_n, O_n^+, Ω_n if $n = 2$.

2) Let $\bar{V}$ be the corresponding projective space. Consider the hypersurface of $\bar{V}$ defined by $X^2 = 0$. What is the geometric meaning of the maximal isotropic subspaces of V?

5. *Special features of symplectic geometry*

We have seen in Theorem 3.7 that every non-singular symplectic space is hyperbolic:

$$\begin{aligned} V = H_{2r} &= P_1 \perp P_2 \perp \cdots \perp P_r \\ &= \langle N_1, M_1, N_2, M_2, \cdots, N_r, M_r \rangle \end{aligned} \tag{3.37}$$

where each $P_i = \langle N_i, M_i \rangle$ is a hyperbolic plane. The N_i, M_j satisfy

(3.38) $$N_iN_j = M_iM_j = 0, \quad N_iM_i = 1, \quad N_\nu M_\mu = 0 \text{ for } \nu \neq \mu.$$

DEFINITION 3.12. We call a basis of V of the type (3.37), (3.38) and in the arrangement (3.37) a symplectic basis of V. The discriminant G of a symplectic basis is 1 as one easily computes.

DEFINITION 3.13. The group of isometries of a non-singular symplectic space $V = H_{2r}$ of dimension $n = 2r$ is called the symplectic group and denoted by $Sp_n(k)$.

We are going to show that $Sp_n(k)$ contains only rotations and shall give two proofs for this fact.

Let A be a given non-zero vector of V. We ask whether $Sp_n(k)$ contains an element σ which moves each vector $X \in V$ by some vector of the line $\langle A \rangle$:

$$\sigma(X) = X + \varphi(X)\cdot A, \qquad \varphi(X) \in k.$$

We should have $\sigma(X)\cdot\sigma(Y) = XY$. This leads to

(3.39) $$\varphi(X)\cdot(AY) = \varphi(Y)\cdot(AX).$$

Select Y_0 so that $(AY_0) \neq 0$, substitute $Y = Y_0$ in (3.39) and compute $\varphi(X)$. We obtain

(3.40) $$\varphi(X) = c\cdot(AX)$$

where $c \in k$ is a constant; (3.40) certainly satisfies (3.39) and consequently

(3.41) $$\sigma(X) = X + c\cdot(AX)\cdot A.$$

We verify immediately that any map of the form (3.41) is a homomorphism of V into V. Since V is non-singular, σ is indeed an isometry. If $c = 0$, then $\sigma = 1_V$. If $c \neq 0$, then σ leaves X fixed if and only if $AX = 0$, which means that X must belong to the hyperplane $H = \langle A \rangle^*$. We can characterize these special isometries in another way. Let H be a given hyperplane. When does σ leave every vector of H fixed? If $X \in V$ and $Y \in H$, then

$$(\sigma X - X)Y = \sigma X\cdot Y - X\cdot Y = \sigma X\cdot Y - \sigma X\cdot\sigma Y$$
$$= \sigma X\cdot(Y - \sigma Y) = \sigma X\cdot 0 = 0$$

since we suppose $\sigma Y = Y$. This means $\sigma X - X \in H^*$. But H^* is a line and we are, therefore, back at our original question.

DEFINITION 3.14. An isometry of the type (3.41) shall be called a symplectic transvection in the direction A.

For a fixed A our σ still depends on c. If we denote it by σ_c we find $\sigma_c \cdot \sigma_d = \sigma_{c+d}$ and $\sigma_c = 1_V$ if and only if $c = 0$. This means that the symplectic transvections in the direction A form a commutative group, isomorphic to the additive group of k.

Let A and B be two non-zero vectors of V. When can we find a symplectic transvection which moves A into B? The direction would have to be $B - A$. Since the case $B = A$ is trivial, we may assume $B \neq A$. Any vector orthogonal to $B - A$ is not moved at all and since A is to be changed we must have $A(B - A) = AB \neq 0$. If this is true, then A is moved by a multiple of $B - A$. Since we still have the c at our disposal we can indeed achieve $\sigma(A) = B$. To repeat: A can be moved into B either if A and B are not orthogonal, or if $A = B$.

Suppose now that $AB = 0$. We can find a C such that $AC \neq 0$ and $BC \neq 0$. Indeed, if $\langle A\rangle^* = \langle B\rangle^*$, select C outside the hyperplane $\langle A\rangle^*$. If $\langle A\rangle^* \neq \langle B\rangle^*$ let $D \in \langle A\rangle^*$, $D \notin \langle B\rangle^*$ and $E \in \langle B\rangle^*$, $E \notin \langle A\rangle^*$; for $C = D + E$ we have $AC = AE \neq 0$ and $BC = BD \neq 0$. Now we can move A into C and then C into B. A non-zero vector A can be moved into some other non-zero vector B by at most two symplectic transvections.

Let now N_1, M_1 and N_2, M_2 be two hyperbolic pairs. By at most two transvections we can move N_1 into N_2 and these transvections will move the hyperbolic pair N_1, M_1 into a pair N_2, M_3. We contend that by at most two more transvections we can move N_2, M_3 into N_2, M_2 and thus have moved N_1, M_1 into N_2, M_2 by at most four symplectic transvections. What we still have to show amounts to the following: if N, M and N, M' are hyperbolic pairs, then we can find at most two transvections which keep N fixed and move M into M'.

Case 1). $MM' \neq 0$. Then we can move M into M' by one transvection whose direction is $M' - M$. Notice that $N(M' - M) = NM' - NM = 1 - 1 = 0$; N remains fixed under this transvection and we are done.

Case 2). $MM' = 0$. We first see that N, $N + M$ is also a hyperbolic pair and $M(N + M) = MN = -1 \neq 0$. Therefore we can move N, M into N, $N + M$. Now we have $(N + M)M' = NM' = 1 \neq 0$ and can, therefore, move N, $N + M$ into N, M'. This finishes the proof.

Let U be a non-singular subspace of V so that

$$V = U \perp U^*.$$

If τ is a transvection of U^*, then $1_U \perp \tau$ is obviously a transvection of V. This shows:

If N_1, M_1 and N_2, M_2 are hyperbolic pairs of U^*, then we can move N_1, M_1 into N_2, M_2 by at most four transvections of V such that each of these transvections keeps every vector of U fixed.

Let now G be the subgroup of $Sp_n(k)$ which is generated by all symplectic transvections. Let $\sigma \in Sp_n(k)$ and N_1, M_1, N_2, M_2, $\cdots$, N_r, M_r be a symplectic basis of V. Denote the images of this basis under σ by N_1', M_1', N_2', M_2', $\cdots$, N_r', M_r'. There is an element $\tau_1 \in G$ which moves N_1, M_1 into N_1', M_1'. Suppose we have constructed an element $\tau_i \in G$ which moves the first i of our hyperbolic pairs N_ν, M_ν into the corresponding N_ν', M_ν'. The images of our first symplectic basis under τ_i will form a symplectic basis

$$N_1', M_1', N_2', M_2', \cdots, N_i', M_i', N_{i+1}'', M_{i+1}'', \cdots, N_r'', M_r''.$$

Let $U = \langle N_1', M_1', N_2', M_2', \cdots, N_i', M_i' \rangle$, then the pairs $N_{i+\nu}'$, $M_{i+\nu}'$ and $N_{i+\nu}''$, $M_{i+\nu}''$ belong all to U^* ($\nu \geq 1$). Find a $\tau \in G$ which keeps every vector of U fixed and moves the pair N_{i+1}'', M_{i+1}'' into the pair N_{i+1}', M_{i+1}'. The element $\tau\tau_i$ will move the first $i + 1$ pairs N_ν, M_ν into the corresponding pairs N_ν', M_ν'. This shows that there is an element of G which moves the whole first basis into the second and we obtain $G = Sp_n(k)$.

Now we will see that $\det \sigma = +1$ for all elements of $Sp_n(k)$. It is indeed enough to show this if σ is a transvection. Nothing is to be proved if the characteristic of k is 2; for, in this case $-1 = +1$. If it is not 2 we can write $\sigma_c = \sigma_{c/2}^2$ and our contention follows.

THEOREM 3.25. *Every element of* $Sp_n(k)$ *is a rotation (i.e., has determinant* $+1$*). The group* $Sp_n(k)$ *is generated by the symplectic transvections.*

If σ is a transvection $\neq 1$ with direction A, then $\sigma X - X \in \langle A \rangle$ for all $X \in V$. If $\tau \in Sp_n(k)$, then

$$\tau\sigma\tau^{-1}X - X = \tau(\sigma(\tau^{-1}X) - (\tau^{-1}X)) \in \langle \tau A \rangle$$

which shows that $\tau\sigma\tau^{-1} \neq 1$ is a transvection with direction τA.

Suppose now that τ belongs to the center of $Sp_n(k)$. Then certainly $\tau\sigma\tau^{-1} = \sigma$ for our transvection and, consequently, $\langle A\rangle = \langle\tau A\rangle$. The element τ leaves all lines of V fixed and Theorem 3.14 implies $\tau = \pm 1_V$.

THEOREM 3.26. *The center of $Sp_n(k)$ consists of $\pm 1_V$. It is of order 2 if the characteristic of k is $\neq 2$, of order 1 if the characteristic is 2.*

The second proof proceeds in an entirely different manner. Let V be a symplectic space, possibly singular. Take any basis of V and compute its discriminant G. If V is singular, then $G = 0$; if V is non-singular, then V has a symplectic basis whose discriminant is $+1$. By formula (3.7) we see that G is the square of an element of k.

Lew now (g_{ij}) be any *skew symmetric* matrix. It induces on a vector space with given basis A_i a symplectic geometry if we define $A_iA_j = g_{ij}$. It follows that the determinant of any skew symmetric matrix with elements $g_{ij} \in k$ is the square of an element of k. Should n be odd, then the symplectic space is singular, the determinant is zero. We shall assume from now on that n is even.

Let Q be the field of rational numbers. Adjoin to Q the $n(n-1)/2$ independent variables x_{ij} $(1 \leq i < j \leq n)$ and call k the resulting field of rational functions of the x_{ij} with rational coefficients. Define x_{ii} to be 0 and put $x_{ij} = -x_{ji}$ for $i > j$. The matrix (x_{ij}) is skew symmetric. Hence

$$\det(x_{ij}) = \left(\frac{f}{g}\right)^2 \tag{3.42}$$

where f and g are polynomials in the x_{ij} $(i < j)$ with integral coefficients. The reader will be familiar with the fact that one has unique factorisation in the ring of polynomials with integral coefficients. We may assume that g is relatively prime to f. The left side of (3.42) is a polynomial with integral coefficients; it follows that f^2 is divisible by g^2 which implies that g is a unit, $g = \pm 1$. The identity (3.42) may now be simplified to

$$\det(x_{ij}) = f^2$$

and f is determined up to a sign. This is a polynomial identity which must remain true in any field and for any (skew symmetric) special values for the x_{ij}. If we specialize the x_{ij} to the g_{ij} of a symplectic basis, then the left side becomes 1. We may, therefore, fix the sign of f by demanding that f should take on the value 1 for a symplectic basis.

THEOREM 3.27. *There exists a polynomial called the Pfaffian and denoted by $Pf(x_{ij})$, which has integral coefficients and the following property: If (g_{ij}) is skew symmetric, then*

$$\det(g_{ij}) = (Pf(g_{ij}))^2.$$

If the g_{ij} come from a symplectic basis, then $Pf(g_{ij}) = 1$.

Adjoin now to Q not only the x_{ij} but n^2 more variables a_{ij} and put

$$(y_{ij}) = (a_{ji})(x_{ij})(a_{ij}).$$

It is easily seen that (y_{ij}) is skew symmetric. Taking determinants on both sides and extracting a square root we find

$$Pf(y_{ij}) = \epsilon \cdot Pf(x_{ij}) \cdot \det(a_{ij})$$

where $\epsilon = \pm 1$. To determine ϵ specialize the (a_{ij}) to the unit matrix when $y_{ij} = x_{ij}$. This shows $\epsilon = +1$.

THEOREM 3.28. *If (g_{ij}) is skew symmetric and (a_{ij}) any matrix, then*

$$Pf((a_{ji})(g_{ij})(a_{ij})) = \det(a_{ij}) \cdot Pf(g_{ij}).$$

Suppose that (a_{ij}) is a matrix describing an element of $Sp_n(k)$ and suppose $Pf(g_{ij}) \neq 0$ which means that the symplectic space is non-singular. Then $(a_{ji})(g_{ij})(a_{ij}) = (g_{ij})$ and we get $\det(a_{ij}) = +1$, another proof of Theorem 3.25.

If V is a non-singular symplectic space of dimension n and (g_{ij}) any skew symmetric matrix, we can always find vectors $A_1, A_2, \cdots, A_n$ in V (not necessarily a basis) such that $g_{ij} = A_iA_j$. This may be proved as follows: Let V_0 be a space with basis $B_1, B_2, \cdots, B_n$. Impose on V_0 a symplectic geometry by putting $B_iB_j = g_{ij}$. This space V_0 may be singular,

$$V_0 = V_1 \perp \text{rad } V_0.$$

Let $B_i = C_i + D_i$ where $C_i \in V_1$, $D_i \in \text{rad } V_0$. Then $g_{ij} = B_iB_j = C_iC_j$. We have $g_{ij} = C_iC_j$ where the vectors C_i come from the non-singular space V_1. Imbed now V_1 in a non-singular space V of the right dimension.

We may take in V a symplectic basis $E_1, E_2, \cdots, E_n$, put $\gamma_{ij} = E_iE_j$ (then $Pf(\gamma_{ij}) = 1$) and set

$$A_j = \sum_{\nu=1}^{n} E_\nu a_{\nu j} .$$

Then

$$g_{ij} = A_iA_j = \sum_{\nu,\mu=1}^{n} \gamma_{\nu\mu}a_{\nu i}a_{\mu j}$$

or

$$(g_{ij}) = (a_{ji})(\gamma_{ij})(a_{ij}).$$

Theorem 3.28 shows

$$Pf(g_{ij}) = \det(a_{ij}). \tag{3.43}$$

From (3.43) one can get all the properties of the Pfaffian which we formulate as exercises.

1) If one interchanges the r-th and s-th row in (g_{ij}) and simultaneously the r-th and s-th column, then $Pf(g_{ij})$ changes the sign.

2) If one multiplies the r-th row *and* the r-th column of (g_{ij}) by t, then $Pf(g_{ij})$ takes on the factor t.

3) $Pf(x_{ij})$ is linear in the row of variables x_{ri} (r fixed, $i = 1, 2, \cdots, n$). Denote by C_{rs} the factor of x_{rs} in $Pf(x_{ij})$ ($r < s$).

4) $$C_{12} = Pf((x_{ij})_{i,j\geq 3})$$

and more generally

$$C_{rs} = Pf((x_{ij})_{i,j\neq r,s})\cdot(-1)^{r+s-1}, \qquad r < s.$$

5) Prove, for instance, the expansion

$$Pf(x_{ij}) = x_{12}C_{12} + x_{13}C_{13} + \cdots + x_{1n}C_{1n} .$$

6) For $n = 2$, $Pf(x_{ij}) = x_{12}$;

for $n = 4$, $Pf(x_{ij}) = x_{12}x_{34} + x_{13}x_{42} + x_{14}x_{23}$.

7) If (a_{ij}) is any matrix of even degree with column vectors A_i and if one defines a "symplectic product" of column vectors by

$$A_iA_j = g_{ij} = \begin{vmatrix} a_{1i} & a_{2i} \\ a_{1j} & a_{2j} \end{vmatrix} + \begin{vmatrix} a_{3i} & a_{4i} \\ a_{3j} & a_{4j} \end{vmatrix} + \cdots + \begin{vmatrix} a_{n-1,i} & a_{ni} \\ a_{n-1,j} & a_{nj} \end{vmatrix}$$

then $\det(a_{ij})$ can be computed by means of

$$\det(a_{ij}) = Pf(g_{ij}).$$

6. *Geometry over finite fields*

Let $k = F_q$ be a field with q elements, F_q^* the multiplicative group of non-zero elements of F_q. Squaring the elements of F_q^* is a homomorphism whose kernel is ± 1. If the characteristic is 2, then $+1 = -1$, the map is an isomorphism, every element is a square. If the characteristic is $\neq 2$, then the kernel is of order 2, the image group has order $(q - 1)/2$ and has index 2 in the whole group F_q^*. If g is a special non-square, then every non-square is of the form gy^2.

Let V be a vector space of dimension n over F_q with a non-singular orthogonal geometry. We investigate first the lowest values of n.

1) $n = 1$, $V = \langle A \rangle$. A^2 may have the form a^2 or the form ga^2; changing A by the factor a we may assume that either $A^2 = 1$ or $A^2 = g$. This gives two distinct geometries: If $A^2 = 1$, then the squares of the vectors $\neq 0$ of V are squares in F_q^*; if $A^2 = g$, then they are non-squares of F_q^*.

2) $n = 2$. A plane over F_q contains $q + 1$ lines, $\langle A + xB \rangle$ and $\langle B \rangle$, if $V = \langle A, B \rangle$. Attach to V a sign $\epsilon = \pm 1$; if V is hyperbolic put $\epsilon = +1$, if V contains no isotropic line put $\epsilon = -1$. The number of non-isotropic lines of V is $q - \epsilon$ (there are just 2 isotropic lines if $\epsilon = +1$). The number of symmetries in $O(V)$ is, therefore, $q - \epsilon$. These symmetries form the coset of $O^+(V)$ in $O(V)$ which is $\neq O^+(V)$. It follows that $O^+(V)$ contains also $q - \epsilon$ elements.

Each of the $q - \epsilon$ non-isotropic lines contains $q - 1$ non-zero vectors. The total number of non-isotropic vectors in V is, therefore, $(q - 1)(q - \epsilon)$. The $q - \epsilon$ rotations produce $q - 1$ equivalence classes among the non-isotropic vectors of V. An equivalence class consists (by Witt's theorem) precisely of the vectors with equal squares. We see that every element of F_q^* must appear as square of some vector of V.

Let A be a vector of V with $A^2 = 1$, B a non-zero vector orthogonal to A; changing B by a factor we may assume that either $B^2 = -1$ or $B^2 = -g$. Since $V = \langle A, B \rangle$ and $(Ax + By)^2 = x^2 + B^2y^2$ we see that V is hyperbolic if $B^2 = -1$ and not hyperbolic if $B^2 = -g$. The sign ϵ really describes the geometry completely; if $\epsilon = -1$, then $x^2 - gy^2$ is the quadratic form attached to our geometry.

For the total number of isotropic vectors in V we find $q + \epsilon q - \epsilon$ ($2q - 1$ if V is hyperbolic, only the zero vector if V is not hyperbolic).

3) $n \geq 3$. Let P be any non-singular plane of V and B a non-isotropic vector of P^*. We can find in P a vector C such that $C^2 = -B^2$. Then $C + B \neq 0$ but $(C + B)^2 = 0$. V must contain isotropic vectors.

According to Theorem 3.11 we can write

$$V = H_{2r} \perp W$$

where H_{2r} is hyperbolic and where W does not contain non-zero isotropic vectors. We see that we have four possible types of geometries (P_i = hyperbolic planes):

n odd

(I) $\quad V = P_1 \perp P_2 \perp \cdots \perp P_{(n-1)/2} \perp \langle A \rangle, \qquad A^2 = 1,$

(II) $\quad V = P_1 \perp P_2 \perp \cdots \perp P_{(n-1)/2} \perp \langle A \rangle, \qquad A^2 = g,$

n even

(III) $\quad V = P_1 \perp P_2 \perp \cdots \perp P_{n/2}\ ,$

(IV) $\quad V = P_1 \perp P_2 \perp \cdots \perp P_{(n-2)/2} \perp W,$

where W is a plane with $\epsilon = -1$.

The quadratic forms attached to these four types are:

(I) $\quad 2x_1x_2 + 2x_3x_4 + \cdots + 2x_{n-2}x_{n-1} + x_n^2\ ,$

(II) $\quad 2x_1x_2 + 2x_3x_4 + \cdots + 2x_{n-2}x_{n-1} + gx_n^2\ ,$

(III) $\quad 2x_1x_2 + 2x_3x_4 + \cdots + 2x_{n-3}x_{n-2} + 2x_{n-1}x_n\ ,$

(IV) $\quad 2x_1x_2 + 2x_3x_4 + \cdots + 2x_{n-3}x_{n-2} + x_{n-1}^2 - gx_n^2\ .$

The discriminants are $(-1)^{(n-1)/2}$, $(-1)^{(n-1)/2}g$, $(-1)^{n/2}$, $(-1)^{n/2}g$. It is worth while noticing that n together with the quadratic character of the discriminant determines the type of geometry. It will also be important to remember that if $V = P_1 \perp V_1$, where P_1 is a hyperbolic plane, then V and V_1 are of the same type.

The difference between type I and type II is inessential. If we multiply the quadratic form of type I by g it becomes equivalent to the form of type II. The difference between type III and type IV is very great; a maximal isotropic subspace is of dimension $n/2$ for type III and of dimension $n/2 - 1$ for type IV.

We associate again a sign $\epsilon = +1$ with type III and $\epsilon = -1$ with type IV.

Denote now by φ_n the number of isotropic vectors of V for some fixed type of geometry. If $n \geq 3$ and N, M is a hyperbolic pair of V, then $\langle N, M\rangle^*$ has dimension $n - 2$ and is of the same type as V; $\langle N, M\rangle^*$ contains, therefore, φ_{n-2} isotropic vectors.

The space $\langle N\rangle \perp \langle N, M\rangle^*$ has dimension $n - 1$ and is orthogonal to N; hence it is the space $\langle N\rangle^*$. Let $xN + A$ be a vector in this space where $A \in \langle N, M\rangle^*$. It will be isotropic if and only if A is isotropic. Thus $\langle N\rangle^*$ contains $q\varphi_{n-2}$ isotropic vectors. To get φ_n we must also find out how many isotropic vectors are *not* orthogonal to N. Such a vector will span a hyperbolic plane with N. A hyperbolic plane containing N is obtained in the form $\langle N, B\rangle$ provided $NB \neq 0$. There are q^n vectors in V and q^{n-1} of them are in $\langle N\rangle^*$ so that we have $q^n - q^{n-1} = q^{n-1}(q - 1)$ such vectors B. Each plane $\langle N, B\rangle$ will contain, by the same reasoning, $q^2 - q = q(q - 1)$ vectors C (not orthogonal to N) which together with N span the same plane $\langle N, B\rangle$. We obtain, therefore $q(q - 1)$ times the same plane if we let B range over the $q^{n-1}(q - 1)$ vectors. Consequently we get q^{n-2} different hyperbolic planes which contain the given N. Each such plane contains exactly one hyperbolic pair N, M whose first vector is the given N. It will contain the isotropic vectors xM (with $x \neq 0$) which are not orthogonal to N; there are $q - 1$ of them in the plane. It follows that there are $q^{n-2}(q - 1) = q^{n-1} - q^{n-2}$ isotropic vectors not orthogonal to N. To summarize:

$q\varphi_{n-2}$ vectors are isotropic and orthogonal to N,

$q^{n-1} - q^{n-2}$ vectors are isotropic and not orthogonal to N,

q^{n-2} is the number of hyperbolic pairs with first component N.

We obtain for φ_n, if $n \geq 3$,

$$\varphi_n = q^{n-1} - q^{n-2} + q\varphi_{n-2}$$

or

$$\varphi_n - q^{n-1} = q(\varphi_{n-2} - q^{n-3})$$

or

$$q^{-n/2}(\varphi_n - q^{n-1}) = q^{-(n-2)/2}(\varphi_{n-2} - q^{n-3}).$$

This means that $q^{-n/2}(\varphi_n - q^{n-1})$ has the same value c for all $n \geq 1$ (and for each type of geometry), in other words,

$$\varphi_n = q^{n-1} + cq^{n/2}, \qquad n \geq 1.$$

For type I or II, n is odd and $\varphi_1 = 1$ (0 is the only isotropic vector if $n = 1$). This shows that $c = 0$, hence

$$\varphi_n = q^{n-1}.$$

For type III or IV, we get for $n = 2$: $\varphi_2 = q + \epsilon q - \epsilon$ so that $c = \epsilon - \epsilon/q$. Therefore, for even n,

$$\varphi_n = q^{n-1} + \epsilon q^{n/2} - \epsilon q^{(n/2)-1} = (q^{n/2} - \epsilon)(q^{(n/2)-1} + \epsilon) + 1.$$

There are $\varphi_n - 1$ isotropic vectors $N \neq 0$. With each of them there are associated q^{n-2} hyperbolic pairs whose first component is N. Thus there are $\lambda_n = q^{n-2}(\varphi_n - 1)$ hyperbolic pairs in V. We have

$$\lambda_n = q^{n-2}(q^{n-1} - 1) \qquad \text{if } n \text{ is odd},$$

$$\lambda_n = q^{n-2}(q^{n/2} - \epsilon)(q^{(n/2)-1} + \epsilon) \qquad \text{if } n \text{ is even}.$$

Before we go on let us figure out the number λ_n of hyperbolic pairs in a *symplectic* case; n must then be even but k may have any characteristic.

Every vector N is isotropic so there are $q^n - 1$ non-zero isotropic vectors. As before, there are $q^n - q^{n-1}$ vectors B not orthogonal to N and $q^2 - q$ of them span the same hyperbolic plane $\langle N, B\rangle$. This gives q^{n-2} hyperbolic planes $\langle N, B\rangle$. In each plane we have $q^2 - q$ vectors B not orthogonal to N and these B give lines $\langle B\rangle$; $q - 1$ among the B give the same line. $\langle N, B\rangle$ contains q lines not orthogonal to N. On each such line one can find one M such that $NM = 1$. We have, therefore, q^{n-1} hyperbolic pairs with first component N. Since there are $q^n - 1$ such N we have

$$\lambda_n = q^{n-1}(q^n - 1)$$

if V is symplectic.

Denote now by Φ_n either the order of O_n^+ for each of our types, or the order of $Sp_n(k)$. A given hyperbolic pair N, M can be moved by a σ into any of the λ_n other hyperbolic pairs. If σ and τ move it into the same pair, then $\tau^{-1}\sigma = \rho$ will leave N, M fixed. Writing $V = \langle N, M\rangle \perp \langle N, M\rangle^*$ we see that $\rho = 1_U \perp \bar{\rho}_{U^*}$ where $U = \langle N, M\rangle$ and where $\bar{\rho}_{U^*}$ is one of the elements of the group of $\langle N, M\rangle^*$ whose order is Φ_{n-2}. Therefore

$$\Phi_n = \lambda_n \Phi_{n-2}$$

where $n \geq 3$ in the orthogonal case, $n \geq 2$ in the symplectic case (we needed the existence of a hyperbolic pair).

If n is odd, then $\Phi_1 = 1$ (identity the only rotation) hence

$$\Phi_n = \lambda_n\lambda_{n-2} \cdots \lambda_3 = q^{(n-2)+(n-4)+\cdots+1} \prod_{i=1}^{(n-1)/2} (q^{2i} - 1).$$

If n is even and V orthogonal, then Φ_2 was determined to $q - \epsilon$ and we have

$$\begin{aligned} \Phi_n &= \lambda_n\lambda_{n-2} \cdots \lambda_4 \cdot (q - \epsilon) \\ &= q^{(n-2)+(n-4)+\cdots+2} \cdot \left(\prod_{i=2}^{n/2} (q^i - \epsilon)(q^{i-1} + \epsilon) \right) (q - \epsilon) \\ &= q^{n(n-2)/4}(q^{n/2} - \epsilon) \cdot \prod_{i=1}^{(n-2)/2} (q^{2i} - 1). \end{aligned}$$

If V is symplectic we have $\Phi_0 = 1$ and get

$$\Phi_n = \lambda_n\lambda_{n-2} \cdots \lambda_2 = q^{(n-1)+(n-3)+\cdots+1} \cdot \prod_{i=1}^{n/2} (q^{2i} - 1).$$

The order Φ_n of our group is, therefore, given by

$$\Phi_n = \begin{cases} q^{(n-1)^2/4} \cdot \prod_{i=1}^{(n-1)/2} (q^{2i} - 1) & \text{if } n \text{ odd, } V \text{ orthogonal} \\ q^{n(n-2)/4}(q^{n/2} - \epsilon) \cdot \prod_{i=1}^{(n-2)/2} (q^{2i} - 1) & \text{if } n \text{ even, } V \text{ orthogonal} \\ q^{(n/2)^2} \cdot \prod_{i=1}^{n/2} (q^{2i} - 1) & \text{if } n \text{ even, } V \text{ symplectic.} \end{cases}$$

Remember that Φ_n denotes the order of O_n^+ in the orthogonal case and *not* the order of O_n.

Consider again the case of an orthogonal geometry. Let U be a space of dimension $r < n$ with a given non-singular geometry. Let W be a space of dimension $n - r$; giving to W the two geometries possible for dimension $n - r$ and forming $U \perp W = V$ we get two possible geometries on V. These geometries are not the same because of Theorem 3.12, since $W = U^*$. This shows that the geometry of W can be selected (uniquely) in such a way that V receives a given one of the two possible geometries. In other words: If V is a given non-singular space of dimension n, then all types of geometries occur among the non-singular proper subspaces of V. One may ask how many subspaces U exist with a given non-singular geometry. Let U be one of them and write

$$V = U \perp U^*.$$

The discriminants of V and U will determine that of U^* and hence the geometry of U^*. If we apply all elements of $O^+(V)$ to U we will get all isometric subspaces. One has only to find how many elements σ of $O^+(V)$ map U onto itself. Then U^* will also be mapped onto itself, hence $\sigma = \tau \perp \rho$ where $\tau \in O(U)$, $\rho \in O(U^*)$. Since σ has to be a rotation $\det \tau = \det \rho$. If we denote by $\Phi(V)$ the order of $O^+(V)$ we get $\Phi(U)\Phi(U^*)$ possibilities where τ and ρ are rotations and just as many where τ and ρ are reflexions. This shows that there are

$$\frac{\Phi(V)}{2\Phi(U)\cdot\Phi(U^*)}$$

subspaces of V, isomorphic to U. Similar questions can be discussed for isotropic subspaces or for given sets of independent vectors ($< n$ in number) which we leave to the reader.

7. *Geometry over ordered fields—Sylvester's theorem*

Suppose that k is an ordered field and V a vector space over k with an orthogonal geometry.

DEFINITION 3.15. The geometry of V is called

a) positive semi-definite if $X^2 \geq 0$ for all $X \in V$,
b) negative semi-definite if $X^2 \leq 0$ for all $X \in V$,
c) positive definite if $X^2 > 0$ for all $X \neq 0$ in V,
d) negative definite if $X^2 < 0$ for all $X \neq 0$ in V.

Suppose V is positive semi-definite and let $Y \in V$ be such that $Y^2 = 0$. For any $X \in V$ and $a \in k$ we have

$$(X + aY)^2 = X^2 + a(XY) \geq 0.$$

If XY were $\neq 0$, then $X^2 + a(XY)$ would range over all of k as a ranges over k and would take on negative values; therefore $XY = 0$

for all $X \varepsilon V$. Hence the radical of V consists of all $Y \varepsilon V$ with $Y^2 = 0$; a positive semi-definite space is non-singular if and only if it is positive definite. The same conclusions hold for negative semi-definite spaces. Observe that the 0-space is positive definite as well as negative definite.

Assume now merely that V is non-singular and let U be a positive definite subspace, W a negative definite subspace. For $X \varepsilon U \cap W$ we conclude $X^2 \geq 0$, $X^2 \leq 0$, hence $X = 0$, $U \cap W = 0$. Therefore $\dim(U + W) = \dim U + \dim W$ and, since $\dim(U + W) \leq n$, we get $\dim U + \dim W \leq n$.

Suppose now that U is not properly contained in any positive definite subspace of V (U is maximal positive definite). We have $V = U \perp U^*$. If U^* would contain a vector Y with $Y^2 > 0$, then $U \perp \langle Y \rangle$ would clearly be positive definite; therefore $Y^2 \leq 0$ for all $Y \varepsilon U^*$. This means, since U^* is non-singular, that U^* is negative definite. We call r the dimension of this particular U. Then we know:

a) Whenever W is negative definite, then $\dim W \leq n - r$.

b) The value $n - r$ is reached, namely for $W = U^*$.

This is an invariant characterisation of the number r and we conclude that *all* maximal positive definite subspaces have the *same* dimension r.

We just need one more piece of information. Suppose U is positive definite. When will it be maximal positive definite? We certainly must have U^* negative definite. Is this sufficient? If U^* is negative definite of dimension $n - s$, then *any* positive definite space has dimension $\leq n - (n - s) = s$. Thus U (which has dimension s) is maximal positive definite (and of course $s = r$).

It is easy to interpret these results in terms of an *orthogonal* basis $A_1, A_2, \cdots, A_n$ of V. Assume $A_1, A_2, \cdots, A_r$ are those basis vectors whose squares are > 0 and let $U = \langle A_1, A_2, \cdots, A_r \rangle$ ($U = 0$ if $r = 0$). Then $U^* = \langle A_{r+1}, A_{r+2}, \cdots, A_n \rangle$ is negative definite and U, therefore, maximal positive definite. Thus the number r is our previous invariant.

The statements we have just made are known as *Sylvester's theorem.*

Our new invariant r does not (in general) describe the geometry of V completely. It does so, however, in one important case. Assume that every positive element of k is a square of an element of k. This holds, for instance, if $k = R$, the field of real numbers. If we replace each basis vector A_i by a suitable $a_i A_i$ we may assume that for the new basis vectors we have $A_i^2 = 1$ for $i \leq r$ and $A_i^2 = -1$ for $i \geq r + 1$. The geometry is based on the quadratic form

$$x_1^2 + x_2^2 + \cdots + x_r^2 - x_{r+1}^2 - \cdots - x_n^2 .$$

We get $n + 1$ such geometries since $r = 0, 1, \cdots, n$ are the possible values for r. The point is that Sylvester's theorem assures us that these geometries are distinct.

CHAPTER IV

The General Linear Group

1. *Non-commutative determinants*

J. Dieudonné has extended the theory of determinants to non-commutative fields. His theory includes the case of ordinary determinants and we shall present it here. Let k be a field which is not necessarily commutative. Let $A = (a_{ij})$ denote an n-by-n matrix with elements in k where the letter i designates the row and j the column of A. For any $i \neq j$ and any $\lambda \in k$, we denote by $B_{ij}(\lambda)$ the matrix obtained from the unit matrix by replacing the element $a_{ij} = 0$ of the unit matrix by λ.

To multiply any matrix A on the left by $B_{ij}(\lambda)$ amounts to adding to the i-th row of A the j-th row multiplied on the left by λ. To multiply A on the right by $B_{ij}(\lambda)$ amounts to adding to the j-th column of A the i-th column multiplied on the right by λ. Specifically, $B_{ij}(\lambda)B_{ij}(\mu) = B_{ij}(\lambda + \mu)$, and $B_{ij}(\lambda)^{-1} = B_{ij}(-\lambda)$.

If $A_1, A_2, \cdots, A_n$ are the row vectors of A, then the row vectors of BA are left linear combinations of the A_ν. A is non-singular if it has an inverse B, i.e., if the n unit vectors can be written as left linear combinations of the A_ν. This is the case if and only if the A_ν are left linearly independent.

The group of all non-singular n-by-n matrices is called the general linear group and denoted by $GL_n(k)$. The matrices $B_{ij}(\lambda)$ (for all $i \neq j$ and all $\lambda \in k$) generate a subgroup, $SL_n(k)$, called the unimodular group. Its elements shall be called unimodular matrices.

Let A be non-singular. We are going to multiply A on the left by various $B_{ij}(\lambda)$ and achieve an especially simple form. We shall then have multiplied A on the left by a certain unimodular matrix B.

Since A is non-singular, not all elements a_{i1} of the first column can be zero. If $n \geq 2$ and $a_{21} = 0$, we add a suitable row to the second row and obtain a matrix with $a_{21} \neq 0$. Now we multiply the second row by $(1 - a_{11})a_{21}^{-1}$ and add it to the first row. This gives a new matrix with $a_{11} = 1$. Now we multiply the first row by a_{i1} and subtract it from the i-th row ($i > 1$). We have then a non-singular matrix with $a_{11} = 1$ and $a_{i1} = 0$ for $i > 1$.

The $n - 1$ rows $A_2, \cdots, A_n$ of this matrix are left linearly independent; thus we can treat the second column in a similar way and obtain $a_{22} = 1$, $a_{2i} = 0$ for $i > 2$ if $n \geq 3$. But we can also achieve $a_{12} = 0$ by subtracting a multiple of the second row from the first.

We are stopped in this procedure only when the matrix differs from the unit matrix just by the last column. The element $a_{nn} = \mu$ must then be $\neq 0$, otherwise the last row would contain only zeros. Therefore, we can at least achieve $a_{in} = 0$ for $i \leq n - 1$, by subtracting multiples of the last row from the others. We end up with a matrix $D(\mu)$, differing from the unit matrix only in the element a_{nn} which is μ. Thus we have

THEOREM 4.1. *Every non-singular matrix A can be written in the form $B \cdot D(\mu)$ with $B \varepsilon SL_n(k)$ and some $\mu \neq 0$.*

To multiply A on the left by $D(\mu)$ amounts to multiplying the last row from the left by μ. To multiply A on the right by $D(\mu)$ amounts to multiplying the last column of A on the right by μ. Specifically,

$$D(\lambda)D(\mu) = D(\lambda\mu).$$

The elements $\neq 0$ of k form a multiplicative group k^*. Its factor commutator group is abelian and shall be denoted by $\overline{k^*}$. To this group, we adjoin a zero element with obvious multiplication, and call the semi-group thus obtained $\bar{k}$. Every $a \neq 0$ of k has a canonical image $\bar{a}$ in $\bar{k}$. With the zero element of k, we associate the zero element $\bar{0}$ of $\bar{k}$. Then $\overline{ab} = \bar{a}\,\bar{b} = \bar{b}\,\bar{a} = \overline{ba}$ and $\bar{1}$ is the unit element of $\bar{k}$. We frequently write simply 1 instead of $\bar{1}$ and 0 instead of $\bar{0}$. Caution is necessary with -1 since it *may* happen (e.g., in the quaternions) that $\overline{-1} = \bar{1}$.

We associate now with every matrix A an element of $\bar{k}$ called determinant: det A. We shall ask that det A satisfies the following axioms:

1) Suppose A' is obtained from A by multiplying one row of A on the left by μ. Then

$$\det A' = \bar{\mu} \det A.$$

2) If A' is obtained from A by adding one row to another, then

$$\det A' = \det A.$$

3) The unit matrix has determinant 1.

Since we are going to prove the existence of determinants by induction on n, we first derive some consequences of the axioms:

a) If we add to the row A_i a left multiple λA_j of another row, the determinant does not change.

Proof: This is obvious if $\lambda = 0$. Let $\lambda \neq 0$, and A' be the matrix obtained from A by replacing A_j by λA_j. Then $\det A = \overline{\lambda^{-1}} \det A'$. We now add the row λA_j of A' to A_i and factor out λ from the j-th row.

b) If A is singular, then $\det A = 0$. Indeed, one row vector is a left linear combination of the others. Subtracting this linear combination from it, we get a matrix with one row $= 0$. Assume A has this form. We factor out 0 from this row and have $\det A = \bar{0} \det A = 0$.

c) If A_i and A_j are interchanged, then $\det A$ is multiplied by $\overline{-1}$.

Proof: Replace A_i by $A_i + A_j$; subtract this from A_j (obtaining $-A_i$) and add this new row to the i-th row. This interchanges A_i and A_j but with a sign change. Factor out $\overline{-1}$.

d) $\det D(\mu) = \bar{\mu}$. Factor out μ from the last row and use axiom 3).

e) If A is non-singular and of the form $B \cdot D(\mu)$ with unimodular B, then $\det A = \bar{\mu}$. This follows from d), since multiplication by B means repeated operations which do not change the determinant.

f) The axioms are categorical. This follows from b) and e).

g) $\det A = 0$ if and only if A is singular. This follows also from b) and e).

h) $\det(AB) = \det A \cdot \det B$.

Proof: If A is singular, AB can not have an inverse, since $ABC = I$ (where I is the unit matrix) implies that $BC = A^{-1}$. Therefore, $\det A = \det AB = 0$, and our formula holds.

If $A = CD(\mu)$ is non-singular where C is unimodular, then $\det A = \bar{\mu}$. The matrix $D(\mu) \cdot B$ is obtained from B by multiplying the last row by μ. Therefore, $\det(D(\mu)B) = \bar{\mu} \det B$. Multiplying $D(\mu)B$ on the left by the unimodular matrix C does not change the determinant. Hence,

$$\det(AB) = \det(D(\mu)B) = \bar{\mu} \det B = \det A \det B.$$

i) Since $\det(A \cdot B_{ij}(\lambda)) = \det A$, it follows that $\det A$ does not change if a right multiple of one column is added to another.

j) If a column of A is multiplied on the right by μ, then the determinant is multiplied by $\bar{\mu}$. This follows for the last column from $\det(AD(\mu)) = \det A \cdot \bar{\mu}$ and for the others by column exchange. The rule for column exchange follows from i) as in c); one obtains it at first only if the last column is interchanged with another. This is enough to prove j) and one can now prove the full rule for column exchange.

Existence of determinants

1) For one-rowed matrices $A = (a)$, we set $\det A = \bar{a}$, and see that the axioms are trivially satisfied. Let us suppose, therefore, that determinants are defined for $(n - 1)$-rowed matrices and that the axioms are verified.

2) If A is singular, put $\det A = 0$. A being singular means that the rows are dependent. Then axioms 1) and 2) are true since the new rows are also dependent.

3) If A is non-singular, then the row vectors A_ν are left linearly independent, and there exist *unique* elements $\lambda_i \in k$ such that $\sum_{\nu=1}^{n} \lambda_\nu A_\nu = (1, 0, \cdots, 0)$ (therefore, not all $\lambda_i = 0$). We write $A_i = (a_{i1}, B_i)$ where B_i is a vector with $n - 1$ components. Then

$$\sum_{\nu=1}^{n} \lambda_\nu a_{\nu 1} = 1 \quad \text{and} \quad \sum_{\nu=1}^{n} \lambda_\nu B_\nu = 0.$$

Consider now the matrix F with n rows B_i (and $n - 1$ columns). Call C_i the $(n - 1)$-by-$(n - 1)$ matrix obtained from F by crossing out the i-th row. We wish to get information about $\det C_i$.

a) If $\lambda_i = 0$, then $\sum_{\nu=1}^{n} \lambda_\nu B_\nu = 0$ shows that the rows of C_i are dependent, and consequently $\det C_i = 0$.

b) Suppose $\lambda_i \neq 0$ and $\lambda_j \neq 0$ $(i \neq j)$. Call D and E the matrices obtained from C_i by replacing B_j by $\lambda_j B_j$ and B_i, respectively.

We have: $\det C_i = \overline{\lambda_j^{-1}} \det D$. If we add to the row $\lambda_j B_j$ all the other rows multiplied on the left by λ_ν, we obtain in this row $\sum_{\nu \neq i} \lambda_\nu B_\nu = -\lambda_i B_i$. Factoring out $-\lambda_i$, we get

$$\det C_i = \overline{\lambda_j^{-1}(-\lambda_i)} \det E.$$

With $|i - j| - 1$ interchanges of adjacent rows, we can change the matrix E into the matrix C_j. Hence we have

$$\det C_i = \overline{(-1)^{i-j-1}\lambda_j^{-1}(-\lambda_i)} \det C_j ,$$

or

$$(4.1) \qquad \overline{(-1)^{i+1}}\ \overline{\lambda_i^{-1}} \det C_i = \overline{(-1)^{j+1}}\ \overline{\lambda_j^{-1}} \det C_j \ .$$

The expression (4.1) is, therefore, the same for any $\lambda_i \neq 0$ and shall be called the determinant of A:

$$\det A = \overline{(-1)^{i+1}}\ \overline{\lambda_i^{-1}} \det C_i \ .$$

Next we have to prove our axioms.

1) Suppose A_i is replaced by μA_i. If $\mu = 0$, the axiom holds trivially. Let $\mu \neq 0$. We have then to replace λ_i by $\lambda_i \mu^{-1}$ and to keep the other λ_ν fixed. If $\lambda_i \neq 0$, then C_i is not changed; the factor $\overline{\lambda_i^{-1}}$ is replaced by $\bar{\mu}\ \overline{\lambda_i^{-1}}$, and the axiom is true. If $\lambda_\nu \neq 0$ for $\nu \neq i$, then $\det C_\nu$ changes by the factor $\bar{\mu}$, and the axiom is again true.

2) If A_i is replaced by $A_i + A_j$, then λ_j is replaced by $\lambda_j - \lambda_i$ and all other λ_ν stay the same; for,

$$\lambda_i(A_i + A_j) + (\lambda_j - \lambda_i)A_j = \lambda_i A_i + \lambda_j A_j \ .$$

If some λ_ν is $\neq 0$ for $\nu \neq i, j$, we use it for the computation of the determinant and see that $\det C_\nu$ does not change since a row is added to another. If $\lambda_i \neq 0$, then λ_i and C_i stay the same. There remains the case when $\lambda_j \neq 0$ but all other $\lambda_\nu = 0$. Then $\lambda_j B_j = 0$; therefore, $B_j = 0$. Since B_i has to be replaced by $B_i + B_j$, no change occurs in C_j; λ_j also does not change, and our axiom is established in all cases.

3) For the unit matrix, we have $\lambda_1 = 1$, all other $\lambda_\nu = 0$. C_1 is again the unit matrix whence $\det I = 1$.

THEOREM 4.2. *Let a, $b \in k$ and $\neq 0$. Let $c = aba^{-1}b^{-1}$. Then $D(c)$ is unimodular if $n \geq 2$.*

Proof: We start with the unit matrix and perform changes which amount always to adding a suitable left multiple of one row to another. The arrow between the matrices indicates the change:

$$\begin{pmatrix} 1 & 0 \\ 0 & 1 \end{pmatrix} \to \begin{pmatrix} 1 & 0 \\ a^{-1} & 1 \end{pmatrix} \to \begin{pmatrix} 0 & -a \\ a^{-1} & 1 \end{pmatrix} \to \begin{pmatrix} 0 & -a \\ a^{-1} & b^{-1} \end{pmatrix}$$

$$\to \begin{pmatrix} aba^{-1} & 0 \\ a^{-1} & b^{-1} \end{pmatrix} \to \begin{pmatrix} aba^{-1} & 0 \\ 1 & b^{-1} \end{pmatrix} \to \begin{pmatrix} 0 & -c \\ 1 & b^{-1} \end{pmatrix}$$

$$\to \begin{pmatrix} 0 & -c \\ 1 & c \end{pmatrix} \to \begin{pmatrix} 1 & 0 \\ 1 & c \end{pmatrix} \to \begin{pmatrix} 1 & 0 \\ 0 & c \end{pmatrix}.$$

The steps are only written out in 2-rowed matrices and are meant to be performed with the last two rows of n-rowed matrices. They show that $D(c)$ is obtained from the unit matrix by multiplication from the left by unimodular matrices, and our theorem is proved.

We can now answer the question as to when $\det A = 1$. If we put $A = BD(\mu)$ with an unimodular B, then $\det A = \bar{\mu}$. But $\bar{\mu} = 1$ means that μ is in the commutator group, therefore a product of commutators. This shows that $D(\mu)$ is unimodular and, hence, A. If $n = 1$, $\det A = \det(a) = \bar{a}$. To get a uniform answer, we may define $SL_1(k)$ to be the commutator group of k^*.

The map $GL_n(k) \to \overline{k^*}$ given by $A \to \det A$ is a homomorphism because of the multiplication theorem for determinants. The map is onto since $D(\mu) \to \bar{\mu}$, and since $\bar{\mu}$ can be any element of $\overline{k^*}$. Its kernel is $SL_n(k)$. $SL_n(k)$ is, consequently, an invariant subgroup of $GL_n(k)$ whose factor group is canonically (by the determinant map) isomorphic to $\overline{k^*}$. Thus we have

THEOREM 4.3. *The formula* $\det A = 1$ *is equivalent with* $A \in SL_n(k)$. $SL_n(k)$ *is an invariant subgroup of* $GL_n(k)$; *it is the kernel of the map* $A \to \det A$, *and the factor group is isomorphic to* $\overline{k^*}$.

THEOREM 4.4. *Let*

$$A = \begin{pmatrix} B & 0 \\ C & D \end{pmatrix},$$

where B and D are square matrices. Then

$$\det A = \det B \cdot \det D.$$

The same is true for

$$A = \begin{pmatrix} B & C \\ 0 & D \end{pmatrix}.$$

Proof:

1) If B is singular, its rows are dependent and, therefore, the corresponding rows of A. Our formula holds.

2) If B is non-singular, we can make unimodular changes (and the same changes on A) to bring it into the form $D(\mu)$. Then $\det B = \bar{\mu}$. Subtracting suitable multiples of the first rows of A from the following, we can make $C = 0$:

$$A = \begin{pmatrix} B & 0 \\ 0 & D \end{pmatrix}.$$

If D is singular, so is A. If D is non-singular, we can assume that D is of the form $D(\rho)$ where $\det D = \bar{\rho}$. Factoring out of A the μ and the ρ, we are left with the unit matrix, and $\det A = \bar{\mu}\bar{\rho} = \det B \cdot \det D$.

For A in the second form, we can argue in the same way starting with D.

Examples for $n = 2$

1) $$\begin{vmatrix} 0 & \beta \\ \gamma & \delta \end{vmatrix} = \overline{-1} \begin{vmatrix} \gamma & \delta \\ 0 & \beta \end{vmatrix} = \overline{-1}\,\overline{\gamma\beta} = \overline{-\gamma\beta}.$$

2) $\alpha \neq 0$; $$\begin{vmatrix} \alpha & \beta \\ \gamma & \delta \end{vmatrix} = \begin{vmatrix} \alpha & \beta \\ 0 & \delta - \gamma\alpha^{-1}\beta \end{vmatrix} = \overline{\alpha\delta - \alpha\gamma\alpha^{-1}\beta}.$$

3) Suppose $ab \neq ba$, then

$$\begin{vmatrix} 1 & a \\ b & ab \end{vmatrix} = \overline{ab - ba} \neq 0.$$

This shows that the right factor b of the last row can not be factored out since otherwise we would get 0.

We shall now investigate the substitutes for the linearity of determinants in the commutative case.

Let k' be the commutator group of the non-zero elements of k. The element $\bar{a}$ can then be interpreted as the coset ak', and we have trivially: $\bar{a}\bar{b} = \bar{a}b = \overline{ab}$. Define addition as follows: $\bar{a} + \bar{b} = ak' + bk'$ is the set of all sums of an element of $\bar{a}$ and an element of $\bar{b}$. Then $a + b \in \bar{a} + \bar{b}$; therefore, $(a + b)\bar{c} \subset a\bar{c} + b\bar{c}$. Notice that such a relation means equality in the commutative case since then $\bar{a} = a$, and $\bar{a} + \bar{b} = a + b$.

Consider now the determinant as a function of some row, say the last one, A_n, keeping the other rows, A_i, fixed. Denote this function by $D(A_n)$.

THEOREM 4.5.

$$D(A_n + A_n') \subset D(A_n) + D(A_n').$$

Proof: The $n + 1$ vectors $A_1, A_2, \cdots, A_n, A'_n$ are linearly dependent: this yields a non-trivial relation

$$\rho_1 A_1 + \cdots + \rho_{n-1} A_{n-1} + \lambda A_n + \mu A'_n = 0.$$

1) If both λ and μ are 0, the first $n - 1$ rows of our three determinants are dependent; the determinants are 0, and our theorem holds.

2) If, say, $\lambda \neq 0$, we can assume $\lambda = 1$. Adding suitable multiples of the first $n - 1$ rows to the last, we find

$$D(A_n) = D(-\mu A'_n) = \overline{-\mu} D(A'_n) = -\mu D(A'_n)$$

and

$$D(A_n + A'_n) = D((1 - \mu)A'_n) = (1 - \mu)D(A'_n).$$

From

$$D(A_n + A'_n) = (1 - \mu)D(A'_n) \subset 1 \cdot D(A'_n) + (-\mu)D(A'_n),$$

we get our theorem.

Corollary. If k is commutative, the determinant is a linear function of each row.

2. *The structure of $GL_n(k)$*

We shall now derive some lemmas about fields which will be necessary later on. Let us denote by

k any field,

Z the center of k which is a commutative subfield of k,

S the *additive* subgroup of k which is generated by the products of squares of k; an element of S is, therefore, of the form $\sum \pm x_1^2 x_2^2 \cdots x_r^2$. Notice that $S + S \subset S$, $S \cdot S \subset S$ but also that $c \in S$ implies $c^{-1} = c \cdot (c^{-1})^2 \in S$ if $c \neq 0$.

Lemma 4.1. *If the square of every element of k lies in Z, then k is commutative.*

Proof:

1) Since $xy + yx = (x + y)^2 - x^2 - y^2$, we know that every element of the form $xy + yx$ belongs to Z.

2) If the characteristic of k is $\neq 2$, we may write for $x \in k$

$$x = \left(\frac{x+1}{2}\right)^2 - \left(\frac{x-1}{2}\right)^2 \tag{4.2}$$

which shows that $x \in Z$.

3) If k has characteristic 2 and is *not* commutative, we can find two elements $a, b \in k$ such that $ab \neq ba$ and thus $c = ab + ba \neq 0$. By 1) we know $c \in Z$ and $ca = a(ba) + (ba)a$ is also in Z. This implies $a \in Z$ and, hence, $ab = ba$ which is a contradiction.

LEMMA 4.2. *Unless k is commutative and of characteristic 2, we have $S = k$.*

Proof:

1) Formula (4.2) proves our lemma if k does not have the characteristic 2.

2) Suppose k has characteristic 2 and is not commutative. Then $xy + yx = (x + y)^2 - x^2 - y^2$ shows $xy + yx \in S$.

By Lemma 4.1 there exists an element $a \in k$ such that $a^2 \notin Z$. We can, therefore, find an element $b \in k$ such that $a^2b \neq ba^2$. The element $c = a^2b + ba^2$ will be $\neq 0$ and in S. Let x be any element of k. Then

$$cx = a^2bx + ba^2x = a^2(bx + xb) + (a^2x)b + b(a^2x)$$

shows that $cx \in S$ and, hence $x \in c^{-1}S \subset S$.

Let V be a right vector space of dimension $n \geq 2$ over k. In Chapter I we have shown that the set $\text{Hom}(V, V)$ of k-linear maps of V into V is isomorphic to the ring of all n-by-n matrices with elements in k. This isomorphism depended on a chosen basis $A_1, A_2, \cdots, A_n$. If $\sigma \in \text{Hom}(V, V)$ and $\sigma A_j = \sum_{\nu=1}^{n} A_\nu a_{\nu j}$, then the elements $a_{\nu j}$ form the j-th column of the matrix (a_{ij}) associated with σ. We define the determinant of σ by

$$\det \sigma = \det(a_{ij}).$$

If the basis of V is changed, then $A = (a_{ij})$ is replaced by some BAB^{-1} and $\det(BAB^{-1}) = \det B \cdot \det A \cdot (\det B)^{-1} = \det A$, since the values of determinants are commutative. The value of $\det \sigma$ does, therefore, not depend on the chosen basis. Clearly $\det(\sigma\tau) = \det \sigma \cdot \det \tau$.

It is preferable to redefine the group $GL_n(k)$ as the set of all non-singular k-linear maps of V into V and the subgroup $SL_n(k)$ by the condition that $\det \sigma = 1$. We shall try to understand the geometric meaning of $SL_n(k)$.

DEFINITION 4.1. An element $\tau \in GL_n(k)$ is called a transvection if it keeps every vector of some hyperplane H fixed and moves any vector $X \in V$ by some vector of $H : \tau X - X \in H$.

We must first determine the form of a transvection. The dual space $\hat{V}$ of V is an n-dimensional left vector space over k. It can be used to describe the hyperplanes of V as follows: if $\varphi \neq 0$ is an element of $\hat{V}$, then the set of all vectors $Y \in V$ which satisfy $\varphi(Y) = 0$ form a hyperplane H of V; if $\psi \in \hat{V}$ describes the same hyperplane, then $\psi = c\varphi$ where $c \in k$.

Suppose our hyperplane H is given by $\varphi \in \hat{V}$. Select a vector B of V which is not in H; i.e., $\varphi(B) = a \neq 0$.

Let X be any vector of V and consider the vector $X - B \cdot a^{-1}\varphi(X)$. Its image under φ is $\varphi(X) - \varphi(B)a^{-1}\varphi(X) = 0$ which implies that it belongs to H. Hence the given transvection τ will not move it:

$$\tau X - \tau B \cdot a^{-1}\varphi(X) = X - B \cdot a^{-1}\varphi(X),$$

$$\tau X = X + (\tau(Ba^{-1}) - Ba^{-1})\varphi(X).$$

Since Ba^{-1} is moved by τ only by a vector of H it follows that $A = \tau(Ba^{-1}) - Ba^{-1} \in H$ and we can write

$$\tau X = X + A \cdot \varphi(X), \qquad \varphi(A) = 0. \tag{4.3}$$

Select conversely any $\varphi \in \hat{V}$ and any $A \in V$ for which $\varphi(A) = 0$; construct the map τ by (4.3) which (in its dependency on A) shall be denoted by τ_A.

If φ or A are zero, then τ is identity, a transvection for any hyperplane. Suppose φ and A are different from 0. We will get $\tau X = X$ if and only if $\varphi(X) = 0$, i.e., if X belongs to the hyperplane H defined by φ. The vector A satisfies $\varphi(A) = 0$ and belongs, therefore, also to H. An arbitrary vector X is moved by a multiple of A, i.e. by a vector of H. The map τ is non-singular since $\tau X = 0$ entails that X is a multiple of A and hence $X \in H$; but then $\tau X = X$ and consequently $X = 0$. We see that (4.3) gives us all transvections; they are much more special than the definition suggests in that a vector X is always moved by a vector of the line $\langle A \rangle \subset H$ which we shall call the direction of the transvection. The notion of direction makes, of course, sense only if the transvection τ is different from 1.

Let $A, B \in H$. We may compute $\tau_A \tau_B$:

$$\begin{aligned}\tau_A(\tau_B X) &= \tau_B X + A\cdot\varphi(\tau_B X)\\ &= \tau_B X + A\cdot(\varphi(X) + \varphi(B)\varphi(X))\\ &= X + B\cdot\varphi(X) + A\cdot\varphi(X)\end{aligned}$$

since $\varphi(B) = 0$. We obtain

$$\tau_A \tau_B = \tau_{A+B} \,.$$

Let $\tau \neq 1$ be the transvection described by (4.3) and σ any element of $GL_n(k)$. Set $\tau' = \sigma\tau\sigma^{-1}$; then

$$\tau'(X) = \sigma\tau_A(\sigma^{-1}X) = \sigma(\sigma^{-1}X + A\cdot\varphi(\sigma^{-1}X)) = X + \sigma A\cdot\varphi(\sigma^{-1}X).$$

The map $X \to \varphi(\sigma^{-1}X) \in k$ is also an element of $\hat{V}$. To find its hyperplane we remark that $\varphi(\sigma^{-1}X) = 0$ is equivalent with $\sigma^{-1}X \in H$, hence with $X \in \sigma H$. The vector σA belongs to σH and we see that $\tau' = \sigma\tau\sigma^{-1}$ is again a transvection which belongs to the hyperplane σH and whose direction is the line $\langle\sigma A\rangle$. Let conversely

$$\tau' X = X + A'\cdot\psi(X)$$

be a transvection $\neq 1$ and H' its hyperplane. We find two vectors B and B' of V which satisfy $\varphi(B) = 1$ and $\psi(B') = 1$.

We determine an element $\sigma \in GL_n(V)$ such that $\sigma A = A'$, $\sigma H = H'$ and $\sigma B = B'$; this can be done since a basis of H together with B will span V and similarly a basis of H' together with B' will also span V. Set $\tau'' = \sigma\tau\sigma^{-1}$. Then

$$\tau''(X) = X + A'\cdot\varphi(\sigma^{-1}X).$$

There is a constant c such that $\varphi(\sigma^{-1}X) = c\psi(X)$ since $\varphi(\sigma^{-1}X)$ also determines the hyperplane $\sigma H = H'$. Setting $X = B'$, $\sigma^{-1}X = B$, we obtain $c = 1$ and find $\tau'' = \tau'$. All transvections $\tau \neq 1$ are conjugate in $GL_n(k)$. From $\tau' = \sigma\tau\sigma^{-1}$ we get also that $\det \tau' = \det \tau$, i.e., that all transvections $\neq 1$ have the same determinant.

We had shown that $\tau_A\tau_B = \tau_{A+B}$, where A and B are vectors of H. Suppose that H contains at least three vectors. Select $A \neq 0$ and then $B \neq 0, -A$. Setting $C = A + B \neq 0$ we have $\tau_A\tau_B = \tau_C$ and all three transvections are $\neq 1$. They have the same determinant $\alpha \neq 0$ and we get $\alpha^2 = \alpha$, $\alpha = 1$. The formula can also be used for another purpose. Let f be the canonical map of $GL_n(k)$ onto its factor commutator group. The image of $GL_n(k)$ under f is a commutative group so that $f(\sigma\tau\sigma^{-1}) = f(\sigma)f(\tau)f(\sigma)^{-1} = f(\tau)$ showing again that all transvections have the same image β under the map f. We again have $\beta^2 = \beta$ and, therefore, $\beta = 1$, the image of a transvection is 1, a transvection lies in the commutator subgroup of $GL_n(k)$. We must remember, however, that we made the assumption that H contains at least three vectors. This is certainly the case if $\dim H \geq 2$ or $n \geq 3$; should $n = 2$, $\dim H = 1$, then it is true if k contains at least three elements. The only exception is, therefore, $GL_2(F_2)$; the result about the determinant holds in this case also since 1 is the only non-zero element of F_2 but the transvections of $GL_2(F_2)$ are not in the commutator group, as we are going to see later.

Let us, for a moment, use a basis $A_1, A_2, \cdots, A_n$ of V. Let τ be the map which corresponds to the matrix $B_{ij}(\lambda)$; τ leaves A_ν fixed if $\nu \neq j$. These A_ν span a hyperplane H. The remaining vector A_j is moved by a multiple of A_i, hence by a vector of H. An arbitrary vector of V is, therefore, also moved by a vector of H; i.e., τ is a transvection. We know that the special transvections $B_{ij}(\lambda)$ generate the group $SL_n(k)$. Thus the group generated by all transvections of V will contain $SL_n(k)$. But we have just seen that every transvection is unimodular, i.e., has determinant 1. It follows that $SL_n(k)$ is the group generated by all transvections of V and this is the geometric interpretation of the group $SL_n(k)$.

Except for $GL_2(F_2)$ we proved that $SL_n(k)$ is contained in the factor commutator group of $GL_n(k)$. The factor group $GL_n(k)/SL_n(k)$ is commutative ($SL_n(k)$ is the kernel of the determinant map) which implies the reverse inclusion. The group $SL_n(k)$ is the commutator group of $GL_n(k)$ and the factor commutator group is canonically isomorphic (by the determinant map) to the factor commutator group of k^*.

If $\tau(X) = X + A\cdot\varphi(X)$ and $\tau'(X) = X + A'\cdot\psi(X)$ are two transvections $\neq 1$ and if B and B' are vectors which satisfy $\varphi(B) = 1$ respectively $\psi(B') = 1$, then select any σ which satisfies $\sigma H = H'$, $\sigma A = A'$ and $\sigma B = B'$ proving that $\tau' = \sigma\tau\sigma^{-1}$.

If $n \geq 3$, we have a great freedom in the construction of σ. The hyperplanes H and H' contain vectors which are independent. Changing the image of a basis vector of H' which is independent of A' by a factor in k we can achieve that $\det \sigma = 1$, i.e., that $\sigma \in SL_n(k)$. All transvections $\neq 1$ are conjugate in $SL_n(k)$.

If $n = 2$, hyperplane and direction coincide. If we insist merely on $\sigma H = H'$, then we can still achieve det $\sigma = 1$, since B' could be changed by a factor. Let now τ range over all transvections $\neq 1$ with hyperplane H. Then $\tau' = \sigma\tau\sigma^{-1}$ will range over transvections with hyperplane H' and any given τ' is obtainable, namely from $\tau = \sigma^{-1}\tau'\sigma$. We see that the whole set of transvections $\neq 1$ with hyperplane H is conjugate, within $SL_n(k)$, to the whole set of transvections $\neq 1$ with some other hyperplane H'.

We collect our results:

THEOREM 4.6. *The transvections $\neq 1$ of V generate the group $SL_n(k)$. They form a single set of conjugate elements in $GL_n(k)$ and, if $n \geq 3$, even a single set of conjugate elements in $SL_n(k)$. If $n = 2$, then the whole set of transvections $\neq 1$ with a given direction is conjugate, within $SL_n(k)$, to the whole set of transvections $\neq 1$ with some other direction.*

THEOREM 4.7. *Unless we are in the case $GL_2(F_2)$ the commutator group of $GL_n(k)$ is $SL_n(k)$ and the factor commutator group is canonically isomorphic (by the determinant map) to the factor commutator group of k^*.*

THEOREM 4.8. *The centralizer of $SL_n(k)$ in $GL_n(k)$ consists of those $\sigma \in GL_n(k)$ which keep all lines of V fixed. They are in one-to-one correspondence with the non-zero elements α of the center Z of k. If σ_α corresponds to $\alpha \in Z^*$, then $\sigma_\alpha(X) = X\alpha$. This centralizer is also the center of $GL_n(k)$. The center of $SL_n(k)$ consists of those σ_α which have determinant 1, for which $\bar{\alpha}^n = 1$.*

Proof: Let $\langle A \rangle$ be a given line and τ a transvection with this direction. If σ is in the centralizer of $SL_n(k)$, then it must commute with τ and $\sigma\tau\sigma^{-1} = \tau$. The left side of this equation has direction $\langle \sigma A \rangle$ the right side $\langle A \rangle$. It follows that $\sigma X = X\alpha$ for any $X \in V$ but α may depend on X. If X_1 and X_2 are independent vectors, then X_1, X_2, $(X_1 + X_2)$ are mapped by σ onto $X_1\alpha$, $X_2\beta$, $(X_1 + X_2)\gamma$. Since $X_1\alpha + X_2\beta$ is also the image of $X_1 + X_2$, we find that $\alpha = \gamma = \beta$. If X_1 and X_2 are dependent, then we compare them to a third independent vector and find again that they take on the same factor. The map $\sigma_\alpha(X) = X\alpha$ is not always linear; it should satisfy $\sigma_\alpha(X\beta) = \sigma_\alpha(X)\beta$ or $X\beta\alpha = X\alpha\beta$ showing $\alpha \in Z^*$. The rest of the theorem is obvious.

We come now to the main topic of this section. We shall call a subgroup G of $GL_n(k)$ invariant under unimodular transformation if $\sigma G \sigma^{-1} \subset G$ for any $\sigma \in SL_n(k)$. We shall prove that such a subgroup G is (with few exceptions) either contained in the center of $GL_n(k)$ or else will contain the whole group $SL_n(k)$. This will be done by showing that G contains all transvections if it is not contained in the center of $GL_n(k)$. The method will frequently make use of the following statement: If $\sigma \in SL_n(k)$ and $\tau \in G$, then $\sigma\tau\sigma^{-1}\tau^{-1}$ is also in G since it is the product of $\sigma\tau\sigma^{-1}$ and of τ^{-1}; $\tau\sigma\tau^{-1}\sigma^{-1}$ and similar expressions lie in G as well.

LEMMA 4.3. *Let G be invariant under unimodular transformations and suppose that G contains a transvection $\tau \neq 1$. We can conclude that $SL_n(k) \subset G$ unless $n = 2$ and—at the same time—k is a commutative field of characteristic 2.*

Proof:

1) Let $n \geq 3$ and let σ range over $SL_n(k)$. The element $\sigma\tau\sigma^{-1}$ ranges over all transvections $\neq 1$ and, consequently, $SL_n(k) \subset G$.

2) Let $n = 2$, $\langle A \rangle$ be the hyperplane (and direction) of τ, and B be some vector independent of A. Then $V = \langle A, B \rangle$ and τ may be described by its effect on A and B:

$$\tau(A) = A, \qquad \tau(B) = A\lambda + B. \tag{4.4}$$

Conversely, any τ of this form is a transvection with hyperplane $\langle A \rangle$. If we wish to denote its dependency on λ, then we shall write τ_λ.

Select $\sigma \in GL_2(V)$ given by

$$\sigma(A) = A\alpha, \qquad \sigma(B) = A\beta + B\gamma. \tag{4.5}$$

Applying σ^{-1} to (4.5) we obtain $A = \sigma^{-1}(A)\alpha$, $\sigma^{-1}(A) = A\alpha^{-1}$ and $B = \sigma^{-1}(A)\beta + \sigma^{-1}(B)\gamma$, hence, $\sigma^{-1}(B) = -A\alpha^{-1}\beta\gamma^{-1} + B\gamma^{-1}$. Obviously $(\tau - 1)A = 0$ and $(\tau - 1)B = A\lambda$, and hence we have $\sigma(\tau - 1)\sigma^{-1}A = 0$ and $\sigma(\tau - 1)\sigma^{-1}B = \sigma(A\cdot\lambda\gamma^{-1}) = A\alpha\lambda\gamma^{-1}$. Therefore

$$\sigma\tau\sigma^{-1}A = A, \qquad \sigma\tau\sigma^{-1}B = A\alpha\lambda\gamma^{-1} + B. \tag{4.6}$$

This computation shall also be used a little later.

If we want σ unimodular, then the determinant of the matrix

$$\begin{pmatrix} \alpha & \beta \\ 0 & \gamma \end{pmatrix},$$

which is $\overline{\alpha\gamma}$, should be $\overline{1}$. This is certainly the case if we set $\gamma = \lambda^{-1}\alpha^{-1}\lambda$. For this choice of γ, (4.6) becomes

$$\sigma\tau\sigma^{-1}A = A, \qquad \sigma\tau\sigma^{-1}B = A\cdot\alpha^2\lambda + B$$

which is the transvection $\tau_{\alpha^2\lambda}$ and lies also in G.

The set T of all λ in k for which $\tau_\lambda \varepsilon G$ contains elements $\neq 0$. If λ_1, $\lambda_2 \varepsilon T$, then $\tau_{\lambda_1}\tau_{\lambda_2}^{\pm 1} = \tau_{\lambda_1 \pm \lambda_2} \varepsilon G$ showing that T is an additive group. We have just seen that for any $\alpha \varepsilon k^*$ the set $\alpha^2 T$ is contained in T. Hence $\alpha_1^2\alpha_2^2 \cdots \alpha_r^2 T \subset T$ and, therefore, $ST \subset T$ where S is the set of Lemma 4.2. By our assumption about k and Lemma 4.2 we conclude that $S = k$ and, consequently, $ST = kT = k$. G contains all transvections with the direction $\langle A \rangle$. Theorem 4.6 shows that G contains all transvections and, consequently, the whole group $SL_2(k)$.

LEMMA 4.4. *Suppose that $n = 2$ and that G contains an element σ whose action on some basis A, B of V is of the type given by formula (4.5). We can conclude that $SL_2(k) \subset G$ if either γ is not in the center of k or, should γ be in the center, if $\alpha \neq \gamma$.*

Proof: Let τ be the transvection given by (4.4). The element $\rho = \sigma\tau\sigma^{-1}\tau^{-1}$ is also in G. For τ^{-1} we find

$$\tau^{-1}(A) = A, \qquad \tau^{-1}(B) = -A\lambda + B.$$

Making use of (4.6) we obtain for the action of ρ:

$$\rho(A) = A, \qquad \rho(B) = A(\alpha\lambda\gamma^{-1} - \lambda) + B.$$

We have a free choice of λ at our disposal and try to select it in such a way that $\alpha\lambda\gamma^{-1} - \lambda \neq 0$; for such a choice, ρ would be a transvection $\neq 1$ of G and Lemma 4.3 would show that $SL_2(k) \subset G$ unless k is a commutative field with characteristic 2. The choice $\lambda = 1$ will work unless $\alpha = \gamma$; by our assumption γ is not in the center if $\alpha = \gamma$ and there exists a λ such that $\gamma\lambda - \lambda\gamma \neq 0$, hence $(\gamma\lambda - \lambda\gamma)\gamma^{-1} = \alpha\lambda\gamma^{-1} - \lambda \neq 0$. There remains the case where k is commutative with characteristic 2. The element γ is then in the center, and, by our assumption, $\alpha \neq \gamma$. The factor is $\alpha\lambda\gamma^{-1} - \lambda = (\alpha\gamma^{-1} - 1)\lambda$ and will range over k if λ ranges over k. The element ρ will then range over *all* transvections with direction $\langle A \rangle$. By Theorem 4.6 all transvections are in G, hence $SL_2(k) \subset G$.

THEOREM 4.9. *Suppose that either $n \geq 3$ or that $n = 2$ but that k contains at least four elements. If G is a subgroup of $GL_n(k)$ which is invariant under unimodular transformations and which is not contained in the center of $GL_n(k)$, then $SL_n(k) \subset G$.*

Proof:

1) $n = 2$. The group G must contain an element σ which moves a certain line $\langle A \rangle$. Set $\sigma A = B$, then V is spanned by A and B and the action of σ may be described in terms of the basis A, B:

$$\sigma A = B, \qquad \sigma B = A\beta + B\gamma, \qquad \beta \neq 0.$$

We select any non-zero element a in k and define an element τ of $GL_2(k)$ by

$$\tau A = A\gamma a - Ba, \qquad \tau B = Aa^{-1}.$$

It belongs to the matrix

$$\begin{pmatrix} \gamma a & a^{-1} \\ -a & 0 \end{pmatrix}$$

which has determinant 1 so that $\tau \in SL_2(k)$. This implies that $\rho = \tau^{-1}\sigma^{-1}\tau\sigma$ is also in G. We find

$$\sigma\tau A = B\gamma a - A\beta a - B\gamma a = -A\beta a, \qquad \sigma\tau B = Ba^{-1}$$

hence, for the inverse of $\sigma\tau$

$$\tau^{-1}\sigma^{-1}A = -Aa^{-1}\beta^{-1}, \qquad \tau^{-1}\sigma^{-1}B = Ba.$$

For $\tau\sigma$ we obtain

$$\tau\sigma A = Aa^{-1}, \qquad \tau\sigma B = A\gamma a\beta - Ba\beta + Aa^{-1}\gamma$$
$$= A(\gamma a\beta + a^{-1}\gamma) - Ba\beta$$

and can now compute $\rho = \tau^{-1}\sigma^{-1}\tau\sigma$:

$$\rho(A) = -Aa^{-1}\beta^{-1}a^{-1},$$
$$\rho(B) = -Aa^{-1}\beta^{-1}(\gamma a\beta + a^{-1}\gamma) - Ba^2\beta.$$

We try to select a in such a way that Lemma 4.4 can be used on ρ. Is a choice of a possible such that $a^2\beta$ is not in the center of k? If β is not in the center, choose $a = 1$. If β is in the center of k, but if k is not commutative, we choose a in such a way that a^2 and, therefore, $a^2\beta$ is not in the center; Lemma 4.1 assures us that this can be done. Thus non-commutative fields are settled. If k is commutative, we try to enforce that $a^2\beta \neq a^{-1}\beta^{-1}a^{-1}$ which is equivalent with $a^4 \neq \beta^{-2}$. The equation $x^4 = \beta^{-2}$ can have at most four solutions; if our field has at least six elements, then we can find an $a \neq 0$ such that $a^4 \neq \beta^{-2}$. This leaves us with just two fields, F_4 and F_5.

In F_4 there are three non-zero elements which form a group, therefore $a^3 = 1$, $a^4 = a$ if $a \neq 0$. All we need is $a \neq \beta^{-2}$ which can, of course, be done.

In F_5 we have $a^4 = 1$ for any $a \neq 0$ and we see that our attempt will fail if $\beta^{-2} = 1$, $\beta = \pm 1$. Should this happen, then we select a in such a way that $a^2\beta = 1$; this is possible since $1^2 = 1$, $2^2 = -1$ in F_5. The action of the corresponding ρ is given by

$$\rho(A) = -A, \qquad \rho(B) = -A\cdot 2\gamma - B.$$

Since $\rho \in G$, we have $\rho^2 \in G$ and

$$\rho^2 A = A, \qquad \rho^2 B = A\cdot 4\gamma + B;$$

if $\gamma \neq 0$, then $\rho^2 \neq 1$ is a transvection and Lemma 4.3 shows $SL_2(k) \subset G$.

This leaves us with the case $k = F_5$, $\beta = \pm 1$, $\gamma = 0$.

If $\beta = 1$, then $\sigma A = B$, $\sigma B = A$. Change the basis: $C = A + B$, $D = A - B$. Then

$$\sigma C = C, \qquad \sigma D = -D.$$

If $\beta = -1$, then $\sigma A = B$, $\sigma B = -A$ and we set $C = A + 2B$, $D = A - 2B$. Then

$$\sigma C = B - 2A = -2C, \qquad \sigma D = B + 2A = 2D.$$

In both cases we can use Lemma 4.4 on σ. This finishes the proof if $n = 2$.

2) $n \geq 3$. Select an element σ in G, which moves a certain line $\langle A \rangle$, and set $\sigma A = B$. Let $\tau \neq 1$ be a transvection with direction $\langle A \rangle$. The element $\rho = \sigma\tau\sigma^{-1}\tau^{-1}$ is again in G and is the product of the transvection τ^{-1} with the direction $\langle A \rangle$ and the transvection $\sigma\tau\sigma^{-1}$ with the direction $\sigma\langle A \rangle = \langle B \rangle$. These directions are distinct, whence $\sigma\tau\sigma^{-1} \neq \tau$, $\rho = \sigma\tau\sigma^{-1}\tau^{-1} \neq 1$.

The transvection τ^{-1} will move a vector X by a multiple of A and the transvection $\sigma\tau\sigma^{-1}$ will move $\tau^{-1}X$ by a multiple of B. It follows that ρ will move X by a vector of the plane $\langle A, B \rangle$. Since $n \geq 3$, we can imbed $\langle A, B \rangle$ in some hyperplane H. Then $\rho H \subset H$, since a vector $X \in H$ is moved by a vector of $\langle A, B \rangle \subset H$. The element $\rho \in G$ has the following properties:

a) $\rho \neq 1$, b) $\rho H = H$, c) $\rho X - X \in H$ for all $X \in V$.

Suppose first that ρ commutes with all transvections belonging to the hyperplane H. Select an element $\varphi \in \hat{V}$ which describes H and let C be any vector of H. Let τ_1 be the transvection

$$\tau_1 X = X + C \cdot \varphi(X).$$

We find $\varphi(\rho X) = \varphi(\rho X - X) + \varphi(X) = \varphi(X)$ because of property c). We obtain

$$\rho\tau_1 X = \rho X + \rho C \cdot \varphi(X),$$
$$\tau_1 \rho X = \rho X + C \cdot \varphi(\rho X) = \rho X + C \cdot \varphi(X).$$

Since we assume right now that $\rho\tau_1 = \tau_1\rho$, a comparison implies that $\rho C = C$ because X can be selected so that $\varphi(X) \neq 0$. But C is an arbitrary vector of H; i.e., the element ρ leaves every vector of H fixed. Together with property c) this implies that ρ is a transvection. By Lemma 4.3 we conclude that $SL_n(k) \subset G$.

We may assume, therefore, that there exists a transvection τ_1 with hyperplane H such that $\rho\tau_1 \neq \tau_1\rho$ and, consequently, $\lambda = \rho\tau_1\rho^{-1}\tau_1^{-1} \neq 1$. The element λ is again in G and is a product of the transvections τ_1^{-1} and $\rho\tau_1\rho^{-1}$ whose hyperplanes are H respectively $\rho H = H$. The element λ is itself a transvection, since it is a product of transvections with the same hyperplane. By Lemma 4.3 we have again $SL_n(k) \subset G$ and our proof is finished.

Theorem 4.9 sheds some light on the definition of determinants. Suppose that we have in some way given a definition of determinants which works for non-singular elements of $\mathrm{Hom}(V,V)$, i.e., for elements of $GL_n(k)$. The only condition that we impose on determinants shall be $\det \sigma\tau = \det \sigma \cdot \det \tau$ but the value of a determinant may be in any (possibly non-commutative) group. This new determinant map is then a homomorphism of $GL_n(k)$ and the kernel G of this map an invariant subgroup of $GL_n(k)$. Let us leave aside the two cases $GL_2(F_2)$ and $GL_2(F_3)$. We would certainly consider a determinant map undesirable if G is contained in the center of $GL_n(k)$, since our map would then amount merely to a factoring by certain diagonal matrices in the center. In any other case we could conclude $SL_n(k) \subset G$ and this implies that the new determinant map is weaker than the old one which has $SL_n(k)$ as the precise kernel.

Definition 4.2. Let Z_0 be the center of the group $SL_n(k)$. The factor group $SL_n(k)/Z_0$ is called the projective unimodular group and is denoted by $PSL_n(k)$.

THEOREM 4.10. *The group $PSL_n(k)$ is simple; i.e., it has no invariant subgroup distinct from* 1 *and from the whole group, except for* $PSL_2(F_2)$ *and* $PSL_2(F_3)$.

Proof: Let Γ be an invariant subgroup of $PSL_n(k)$. Its elements are cosets of Z_0. Let G be the union of these cosets; G is an invariant subgroup of $SL_n(k)$. If $G \subset Z_0$, then $\Gamma = 1$; otherwise G will be $SL_n(k)$ and hence $\Gamma = PSL_n(k)$.

3. *Vector spaces over finite fields*

Suppose that $k = F_q$, the field with q elements. Our groups become finite groups and we shall compute their orders.

Let $A_1, A_2, \cdots, A_n$ be a basis of V and $\sigma \varepsilon GL_n(k)$. The element σ is completely determined by the images $B_i = \sigma A_i$ of the basis vectors. Since σ is non-singular, the B_i must be linearly independent. For B_1 we can select any of the $q^n - 1$ non-zero vectors of V. Suppose that we have already chosen $B_1, B_2, \cdots, B_i$ $(i < n)$; these vectors span a subspace of dimension i which will contain q^i vectors. For B_{i+1} we can select any of the $q^n - q^i$ vectors outside of this subspace. We find, therefore,

$$(q^n - 1)(q^n - q) \cdots (q^n - q^{n-1}) = q^{n(n-1)/2} \prod_{i=1}^{n} (q^i - 1)$$

possibilities for the choice of the images. This number is the order of $GL_n(k)$.

The group $SL_n(k)$ is the kernel of the determinant map of $GL_n(k)$ onto the group $\overline{k^*} = k^*$ (k is commutative by Theorem 1.14) and k^* has $q - 1$ elements. The order of $SL_n(k)$ is, therefore, obtained by dividing by $q - 1$, hence simply by omitting the term $i = 1$ in the product.

The order of the center of $SL_n(k)$ is equal to the number of solutions of the equation $\alpha^n = 1$ in k^*. The elements of k^* form a group of order $q - 1$, they all satisfy $x^{q-1} = 1$. Let d be the greatest common divisor of n and $q - 1$. Then we can find integers r and s such that $d = nr + (q - 1)s$.

If $\alpha^n = 1$, then $\alpha^d = (\alpha^n)^r \cdot (\alpha^{q-1})^s = 1$; conversely, $\alpha^d = 1$ implies $\alpha^n = 1$, since d divides n. We must now find the number of solutions of the equation $\alpha^d = 1$ (where d divides $q - 1$). The equation $x^{q-1} - 1$ has $q - 1$ distinct roots in k, namely all elements of k^*.

It factors into $q - 1$ distinct linear factors. The polynomial $x^d - 1$ divides $x^{q-1} - 1$ and this implies that $x^d - 1$ factors into d distinct linear terms. The order of the center of $SL_n(k)$ is, therefore, d.

To get the order of $PSL_n(k)$ we have to divide the order of $SL_n(k)$ by d.

THEOREM 4.11. *The order of $GL_n(k)$ is $q^{n(n-1)/2} \prod_{i=1}^{n} (q^i - 1)$, that of $SL_n(k)$ is $q^{n(n-1)/2} \prod_{i=2}^{n} (q^i - 1)$ and the order of $PSL_n(k)$ is $(1/d)\, q^{n(n-1)/2} \prod_{i=2}^{n-1} (q^i - 1)$ where d is the greatest common divisor of n and $q - 1$. We know that $PSL_n(k)$ is simple unless $n = 2$, $q = 2$ or $n = 2$, $q = 3$.*

For $n = 2$ we obtain as orders of $PSL_n(k)$:

$$\frac{q(q^2 - 1)}{2} \qquad \text{if } q \text{ is odd,}$$

$$q(q^2 - 1) \qquad \text{if } q \text{ is even.}$$

For $q = 2, 3, 4, 5, 7, 8, 9, 11, 13$ we find as orders 6, 12, 60, 60, 168, 504, 360, 660, 1092.

If V is of dimension n over F_q, then V contains $q^n - 1$ non-zero vectors. On each line there are $q - 1$ non-zero vectors, so that V contains $(q^n - 1)/(q - 1)$ lines. An element of $SL_n(F_q)$ may be mapped onto the permutation of these lines. The kernel (if the lines are kept fixed) is the center of $SL_n(F_q)$. This induces a map of $PSL_n(F_q)$ onto the permutations of the lines which now is an isomorphism into.

If $n = 2$, $q = 2, 3, 4$ we have 3, 4, 5 lines respectively. We see:

1) $GL_2(F_2) = SL_2(F_2) = PSL_2(F_2)$ is the symmetric group of 3 letters. It is not simple and $SL_2(F_2)$ is not the commutator group of $GL_2(F_2)$ which is cyclic of order 3.

2) $PSL_2(F_3)$ is the alternating group of 4 letters. It is not simple either.

3) $PSL_2(F_4)$ is the alternating group of 5 letters. The group $PSL_2(F_5)$ has also order 60 and it can be shown to be isomorphic to $PSL_2(F_4)$.

Other interesting cases:

4) $PSL_2(F_7)$ and $PSL_3(F_2)$have both order 168. They are, in fact, isomorphic.

5) $PSL_2(F_9)$ has order 360 and is isomorphic to the alternating group of 6 letters.

6) The two simple groups $PSL_3(F_4)$ and $PSL_4(F_2)$ have both the order 20160 which is also the order of the alternating group A_8 of 8 letters. One can show that $A_8 \simeq PSL_4(F_2)$. Since it is very interesting to have an example of simple groups which have the same order and are *not* isomorphic, we will show that $PSL_3(F_4)$ and $PSL_4(F_2)$ are not isomorphic.

The non-zero elements of F_4 form a cyclic group of order 3. It follows that the cube of each element in the center of $GL_3(F_4)$ is 1. Take an element $\bar{\sigma}$ of $PSL_3(F_4)$ which is of order 2; $\bar{\sigma}$ is a coset and let σ be an element of $SL_3(F_4)$ in this coset. $\bar{\sigma}^2 = 1$ means σ^2 is in the center of $SL_3(F_4)$, therefore $\sigma^6 = 1$. The element σ^3 lies in the coset $\bar{\sigma}^3 = \bar{\sigma}$ and we could have taken σ^3 instead of σ. Making this replacement we see that the new σ satisfies $\sigma^2 = 1$.

Let H be the kernel of the map $(1 + \sigma)X = X + \sigma X$. Since $V/H \simeq (1 + \sigma)V$, we obtain

$$3 = \dim V = \dim H + \dim(1 + \sigma)V;$$

$\dim H = 3$ would imply $(1 + \sigma)V = 0$, $X + \sigma X = 0$ for all $X \varepsilon V$ hence $\sigma X = X$ (the characteristic is 2). But this would mean $\sigma = 1$ whereas we assumed σ of order 2. Consequently $\dim H \leq 2$. $X \varepsilon H$ means $X + \sigma X = 0$ or $\sigma X = X$; H consists of the elements left fixed by σ. The elements in $(1 + \sigma)V$ are left fixed, since $\sigma(X + \sigma X) = \sigma X + \sigma^2 X = \sigma X + X$ which implies $(1 + \sigma)V \subset H$ and consequently $\dim(1 + \sigma)V \leq \dim H$. This leaves $\dim H = 2$ as the only possibility, H is a hyperplane and σ moves any X by $\sigma X + X \varepsilon (1 + \sigma)V \subset H$; σ is, therefore, a transvection $\neq 1$. Since $n = 3$, any two such transvections $\neq 1$ are conjugate and this shows that the elements of order 2 in $PSL_3(F_4)$ form a single set of conjugate elements.

Let us look at $PSL_4(F_2) = SL_4(F_2) = GL_4(F_2)$. Among the elements of order two we have first of all the transvections $\neq 1$ $(B_{12}(1)^2 = B_{12}(0) = 1)$ which form one class of conjugate elements. Let A_1, A_2, A_3, A_4 be a basis of V and define τ by

$$\tau(A_1) = A_1, \qquad \tau(A_2) = A_1 + A_2,$$

$$\tau(A_3) = A_3, \qquad \tau(A_4) = A_3 + A_4.$$

We have $\tau^2 = 1$ but τ is not a transvection since A_2 and A_4 are moved by vectors of different lines. Thus the elements of order 2 form at least two sets of conjugate elements and $PSL_3(F_4)$ and $PSL_4(F_2)$ are not isomorphic.

It can be shown that no other order coincidences among the groups $PSL_n(F_q)$ or with alternating groups occur aside from the ones we have mentioned.

CHAPTER V

The Structure of Symplectic and Orthogonal Groups

In this chapter we investigate questions similar to those of the preceding chapter. We consider a space with either a symplectic or an orthogonal geometry and try to determine invariant subgroups of the symplectic respectively orthogonal group. This is easy enough in the symplectic case and one may also expect a similar result for the orthogonal groups. However it turns out that not all orthogonal groups lead to simple groups. The problem of the structure of the orthogonal groups is only partly solved and many interesting questions remain still undecided. We begin with the easy case, the one of the symplectic group.

1. Structure of the symplectic group

We shall prove the analogue to Theorem 4.4.

THEOREM 5.1. *Let G be an invariant subgroup of $Sp_n(k)$ which is not contained in the center of $Sp_n(k)$. Then $G = Sp_n(k)$ except for the cases $Sp_2(F_2)$, $Sp_2(F_3)$ and $Sp_4(F_2)$.*

Proof:

1) Let $A \in V$, $A \neq 0$ and suppose that G contains all transvections in the direction A. If B is another non-zero vector, select $\tau \in Sp_n(k)$ such that $\tau A = B$. If σ ranges over the transvections in the direction A, then $\tau\sigma\tau^{-1}$ ranges over the transvections in the direction B so that G contains all transvections. We have $G = Sp_n(k)$ in this case. If $k = F_2$, there is only one transvection $\neq 1$ with direction A and if $k = F_3$, there are two, but they are inverses of each other. In the cases $k = F_2$ or $k = F_3$ it is, therefore, sufficient to ascertain that G contains one transvection $\neq 1$.

2) If $n = 2$, $V = \langle N, M\rangle$ where N, M is a hyperbolic pair, let $\sigma N = \alpha N + \beta M$, $\sigma M = \gamma N + \delta M$. The only condition σ must

satisfy to be in $Sp_2(k)$ is $\sigma N \cdot \sigma M = N \cdot M = 1$ which implies $\alpha\delta - \beta\gamma = 1$. This shows that $Sp_2(k) = SL_2(k)$, the two-dimensional unimodular group of V. Theorem 4.9 proves our contention if $n = 2$.

3) G must contain some σ which moves at least one line $\langle A \rangle$. Let $\sigma A = B$ and assume at first that $AB = 0$. Since $\langle A \rangle \neq \langle B \rangle$, we have $\langle A \rangle^* \neq \langle B \rangle^*$. The hyperplane $\langle B \rangle^*$ is, therefore, not contained in $\langle A \rangle^*$ and we can find a vector $C \in \langle B \rangle^*$, $C \notin \langle A \rangle^*$. Then $CB = 0$ but $CA \neq 0$; we can change C by a factor to make $CA = 1$. Let τ be the transvection

$$\tau X = X + ((C - A)X) \cdot (C - A).$$

Then $\tau A = A + (C - A) = C$ and $\tau B = B$.

The element $\rho = \tau\sigma^{-1}\tau^{-1}\sigma$ is in G and we find $\rho(A) = \tau\sigma^{-1}\tau^{-1}(B) = \tau\sigma^{-1}(B) = \tau(A) = C$. Since $AC \neq 0$, we see that it is no loss of generality to assume of our original σ that $\sigma(A) = B$ and $AB \neq 0$.

Let τ_1 be the transvection $\tau_1 X = X + (AX) \cdot A$. The element $\rho_1 = \tau_1\sigma\tau_1^{-1}\sigma^{-1}$ belongs to G and we obtain

$$\rho_1(B) = \tau_1\sigma\tau_1^{-1}(A) = \tau_1\sigma(A) = \tau_1 B = B + (AB)A. \tag{5.1}$$

The transvection τ_1 leaves every vector orthogonal to A and the transvection $\sigma\tau_1^{-1}\sigma^{-1}$ every vector orthogonal to B fixed. The product ρ_1 will keep every vector orthogonal to the plane $P = \langle A, B \rangle$ fixed. This plane is non-singular, since $AB \neq 0$; we can write $V = P \perp P^*$ and find that $\rho_1 = \rho_2 \perp 1_{P^*}$ where $\rho_2 \in Sp_2(k)$. Equation (5.1) shows that ρ_2 does not leave the line $\langle B \rangle$ of the plane P fixed, i.e., that ρ_2 is not in the center of the group $Sp_2(k)$. The set of all $\lambda \in Sp_2(k)$ for which $\lambda \perp 1_{P^*}$ belongs to G is an invariant subgroup of $Sp_2(k)$ which is not in the center of $Sp_2(k)$. If $k \neq F_2$, F_3 this set will, therefore, be all of $Sp_2(k)$. In particular, if we take for λ all transvections of P in the direction of $\langle A \rangle$, we see that G contains all transvections $\lambda \perp 1_{P^*}$ of V in the direction of $\langle A \rangle$. This proves that $G = Sp_n(k)$.

If $k = F_3$, we have $n \geq 4$. We imbed P in a non-singular space V_4 of dimension 4, the element ρ_2 has the form $\rho_3 \perp 1_{V_4^*}$ where $\rho_3 \in Sp_4(k)$ and we could argue as in the preceding case, provided our theorem is proved for the group $Sp_4(k)$. Similarly, if $k = F_2$, it suffices to prove the theorem for the group $Sp_6(k)$.

We must, therefore, consider the two special cases $Sp_4(F_3)$ and $Sp_6(F_2)$.

4) Suppose $k = F_2$ or F_3, and assume first that the element ρ_1 leaves some line $\langle C \rangle$ of the plane P fixed. If $k = F_2$, then the vector C itself is kept fixed, ρ_1 is a transvection of P, hence of V, and our theorem is true in this case. If $k = F_3$, then ρ_1 may reverse C. Setting $P = \langle C, D \rangle$, we obtain

$$\rho_1 C = -C, \qquad \rho_1 D = -D + \zeta C$$

since the effect of ρ_1 must be unimodular. The element ζ is not 0 since ρ_1 moves the line $\langle A \rangle$ of P. For ρ_1^2 we obtain

$$\rho_1^2 C = C, \qquad \rho_1^2 D = D - 2\zeta C$$

and this is a transvection.

We may assume that ρ_1 keeps no line of P fixed. If $k = F_2$, then there are three non-zero vectors in P and ρ_1 permutes them cyclically; ρ_1^{-1} will give the other cyclic permutation. If Q is another non-singular plane and $\tau \in Sp_6(F_2)$ such that $\tau P = Q$, then the elements $\tau \rho_1^{\pm 1} \tau^{-1}$ are also in G and permute the non-zero vectors of Q cyclically, leaving every vector orthogonal to Q fixed.

If $k = F_3$, let $\langle A \rangle$ be any line of P. Setting $\rho_1 A = B$ we have $P = \langle A, B \rangle$ and the unimodular effect of ρ_1 on P is given by

$$\rho_1 A = B, \qquad \rho_1 B = -A + \beta B.$$

We contend that $\beta = 0$; otherwise we would have $\rho_1(A + \beta B) = B - \beta A + \beta^2 B = -\beta A - B = -\beta(A + \beta B)$ (remember $\beta = \pm 1$ in F_3) showing that the line $\langle A + \beta B \rangle$ is kept fixed. We see that

$$\rho_1 A = B, \qquad \rho_1 B = -A, \qquad \rho_1^3 A = -B, \qquad \rho_1^3 B = A.$$

Since $AB = \pm 1$, one of the pairs A, B and B, A is hyperbolic. The group G contains, therefore, an element σ which has the following effect on a symplectic basis N_1, M_1, N_2, M_2 of V:

$$\sigma N_1 = M_1, \qquad \sigma M_1 = -N_1, \qquad \sigma N_2 = N_2, \qquad \sigma M_2 = M_2.$$

We shall now investigate $Sp_4(F_3)$ and $Sp_6(F_2)$ separately. The following type of argument will be used repeatedly: If $A_1, A_2, \cdots, A_n$ and $B_1, B_2, \cdots, B_n$ are two *symplectic bases* of V, then there exists a $\tau \in Sp_n(k)$ such that $\tau A_i = B_i$. If we know that $\sigma \in G$ and that $\sigma A_i = \sum_{\nu=1}^{n} a_{i\nu} A_\nu$, then $\tau\sigma\tau^{-1}$ is also in G and

$$\tau\sigma\tau^{-1} B_i = \tau\sigma A_i = \sum_{\nu=1}^{n} a_{i\nu} B_\nu .$$

5) $Sp_4(F_3)$; let $N_1, M_1; N_2, M_2$ be a symplectic basis of V, then

$$N_1, N_2 + M_1; \quad N_2, M_2 + N_1 \tag{5.2}$$

is also a symplectic basis. The group G contains an element σ which has the following effect:

$$\sigma N_1 = M_1, \quad \sigma M_1 = -N_1, \quad \sigma N_2 = N_2, \quad \sigma M_2 = M_2. \tag{5.3}$$

It also contains an element τ having a corresponding effect on the basis (5.2):

$$\begin{aligned} \tau N_1 = N_2 + M_1, \quad \tau(N_2 + M_1) = -N_1, \quad \tau N_2 = N_2, \\ \tau(M_2 + N_1) = M_2 + N_1. \end{aligned} \tag{5.4}$$

We find, for $\sigma\tau$,

$$\begin{aligned} \sigma\tau N_1 = N_2 - N_1, \quad \sigma\tau(N_2 + M_1) = -M_1, \quad \sigma\tau N_2 = N_2, \\ \sigma\tau(M_2 + N_1) = M_2 + M_1, \end{aligned}$$

$$\begin{aligned} \sigma\tau N_1 = N_2 - N_1, \quad \sigma\tau M_1 = -M_1 - N_2, \quad \sigma\tau N_2 = N_2, \\ \sigma\tau M_2 = N_1 - N_2 + M_1 + M_2. \end{aligned}$$

The element $\rho = (\sigma\tau)^2$ is also in G and has the form

$$\rho N_1 = N_1, \quad \rho M_1 = M_1, \quad \rho N_2 = N_2, \quad \rho M_2 = N_2 + M_2.$$

But this is a transvection and the proof is finished.

6) $Sp_6(F_2)$; we are going to use the following symplectic bases:

$$N_1, M_1; \quad N_2, M_2; \quad N_3, M_3, \tag{5.5}$$

$$N_1, M_1 + M_2; \quad N_1 + N_2, M_2; \quad N_3, M_3, \tag{5.6}$$

$$N_1, M_1; \quad N_3, M_3; \quad N_2, M_2, \tag{5.7}$$

$$N_1, M_1; \quad N_2 + N_3, M_2; \quad N_3, M_2 + M_3. \tag{5.8}$$

Let σ_1 be the cyclic interchange (on the basis (5.5))

$$N_2 \to N_2 + M_2 \to M_2 \to N_2$$

keeping N_1, M_1, N_3, M_3 fixed.

Similarly, let σ_2 be the change (on the basis (5.6))

$$N_1 + N_2 \to M_2 \to N_1 + N_2 + M_2 \to N_1 + N_2$$

keeping N_1, $M_1 + M_2$, N_3, M_3 fixed.

Set $\sigma_3 = \sigma_2\sigma_1$. We find

$$\sigma_3 M_1 = M_1 + N_1 + N_2, \qquad \sigma_3 M_2 = M_2 + N_1, \qquad \sigma_3 M_3 = M_3$$

and the vectors N_i are kept fixed.

Call σ_4, σ_5 the elements which have an effect similar to σ_3 on the basis (5.7), respectively (5.8). They keep the vectors N_i fixed and we obtain

$$\begin{aligned} \sigma_4 M_1 &= M_1 + N_1 + N_3, & \sigma_4 M_2 &= M_2, \\ \sigma_4 M_3 &= M_3 + N_1, & \sigma_5 M_1 &= M_1 + N_1 + N_2 + N_3, \\ \sigma_5 M_2 &= M_2 + N_1, & \sigma_5 M_3 &= M_3 + N_1. \end{aligned}$$

The element $\tau = \sigma_3\sigma_4\sigma_5$ lies also in G, keeps the vectors N_i fixed and since

$$\tau M_1 = M_1 + N_1, \qquad \tau M_2 = M_2, \qquad \tau M_3 = M_3,$$

it is a transvection. This completes the proof.

DEFINITION 5.1. The factor group of $Sp_n(k)$ modulo its center is called the projective symplectic group and denoted by $PSp_n(k)$.

THEOREM 5.2. *The groups $PSp_n(k)$ are simple except for $PSp_2(F_2)$, $PSp_2(F_3)$ and $PSp_4(F_2)$.*

The center of $Sp_n(k)$ is identity if the characteristic of k is 2, it is of order 2 if the characteristic is $\neq 2$. In Chapter III, §6, we have determined the order of $Sp_n(F_q)$. If we denote by d the greatest common divisor of 2 and $q - 1$, then we find, as order of $PSp_n(F_q)$,

$$\frac{1}{d} q^{(n/2)^2} \prod_{i=1}^{n/2} (q^{2i} - 1).$$

The order of the simple group $PSp_4(F_3)$ is 25920. This group turns up in algebraic geometry as the group of the 27 straight lines on a cubic surface.

The order of the simple group $PSp_6(F_2)$ is 1451520. It also arises in algebraic geometry as the group of the 28 double tangents to a plane curve of degree four.

2. *The orthogonal group of euclidean space*

We specialize k to the field R of real numbers. Let V be an n-dimensional vector space over R with an orthogonal geometry based on the quadratic form

$$x_1^2 + x_2^2 + \cdots + x_n^2 .$$

Then we say that V is a euclidean space. Let us first look at the case $n = 3$. No isotropic vectors are present, the geometry on each plane P of V is again euclidean so that any two planes P_1 and P_2 of V are isometric. $P_2 = \tau P_1$ for some $\tau \in O_3^+$. It follows that all 180° rotations are conjugate. Since they generate O_3^+ we can say: If an invariant subgroup G of O_3^+ contains one 180° rotation, then $G = O_3^+$.

Now let $G \neq 1$ be an invariant subgroup of O_3^+ and $\sigma \neq 1$ an element of G. We look at the action of G on the unit sphere of vectors with square 1; σ has an axis $\langle A \rangle$ where $A^2 = 1$. In geometric language the argument is very simple: σ rotates the sphere around the axis $\langle A \rangle$ by a certain angle. Let d be the distance by which a vector Y on the equator ($AY = 0$) is moved. The other vectors of the sphere are moved by at most d, and to a given number $m \leq d$ one can find a point Z_1 which is moved by the distance m (into a point Z_2). If P_1, P_2 are any points on the sphere with distance m, then we can find a rotation $\tau \in O_3^+$ such that $\tau Z_1 = P_1$, $\tau Z_2 = P_2$. The rotation $\tau\sigma\tau^{-1}$ will also be in G and will move P_1 into P_2. This shows that a point P_1 can be moved by an element of G into any point P_2 provided the distance of P_1 and P_2 is $\leq d$. By repeated application of such motions one can now move P_1 into any point of the sphere, especially into $-P_1$. Hence G contains an element σ_1 which reverses a certain vector. But in Chapter III—when we discussed the case $n = 3$—we saw that σ_1 must be a 180° rotation. Thus $G = O_3^+$.

Some readers might like to see the formal argument about the point Z_1: set $Z_1 = A \cdot \sqrt{1 - \mu^2} + Y \cdot \mu$ where $0 \leq \mu \leq 1$. Then

$$(\sigma - 1)Z_1 = \mu \cdot (\sigma - 1)Y, \qquad ((\sigma - 1)Z_1)^2 = \mu^2((\sigma - 1)Y)^2$$

which makes the statement obvious.

The group O_4^+ modulo its center is not simple, but we can show

THEOREM 5.3. *Let V be a euclidean space of dimension $n \geq 3$, but $n \neq 4$. If G is an invariant subgroup of O_n^+ which is not contained* *in the center of* O_n^+ , *then* $G = O_n^+$. *This means that* O_n^+ *is simple for odd dimensions* ≥ 3; *for even dimensions* ≥ 6 *the factor group of* O_n^+ *modulo its center* $\pm 1_V$ *is simple.*

Proof: We may assume $n \geq 5$. Let σ be an element of G which is not in the center of $O(V)$. If we had $\sigma P = P$ for every plane P of V, then we would also have $\sigma L = L$ for every line L of V, since L is the intersection of planes. Let P be a plane such that $\sigma P \neq P$ and λ the 180° rotation in the plane P. Then $\sigma\lambda\sigma^{-1}$ is the 180° rotation in the plane σP and consequently $\neq \lambda$. The element $\rho = \sigma\lambda\sigma^{-1}\lambda^{-1}$ is also in G, is $\neq 1$, and will leave every vector orthogonal to $P + \sigma P$ fixed. Since $n \geq 5$, a non-zero vector A exists which is orthogonal to $P + \sigma P$. Since $\rho A = A$, we have $\rho \neq -1_V$; since $\rho \neq 1_V$, we know that ρ is not in the center of $O(V)$.

The element ρ must move some line $\langle B \rangle$; set $\rho B = C$. Denote by τ_A the symmetry with respect to the hyperplane $\langle A \rangle^*$. The product $\mu = \tau_B \tau_A$ is a rotation. We have $\rho\tau_A\rho^{-1} = \tau_{\rho A} = \tau_A$ and $\rho\tau_B\rho^{-1} = \tau_C$ and, hence, $\rho\mu\rho^{-1}\mu^{-1} = \tau_C\tau_A\tau_A^{-1}\tau_B^{-1} = \tau_C\tau_B$; the element $\sigma_1 = \tau_C\tau_B$ lies in G, is $\neq 1$ since $\langle C \rangle \neq \langle B \rangle$ and leaves every vector orthogonal to the plane $Q = \langle B, C \rangle$ fixed. If U is a three-dimensional subspace of V which contains Q, we may write $\sigma_1 = \rho_1 \perp 1_{U^*}$ where $\rho_1 \in O_3^+(U)$ and $\rho_1 \neq 1$. The group $O_3^+(U)$ is simple and we conclude that G contains all elements of the form $\rho_2 \perp 1_{U^*}$ where $\rho_2 \in O_3^+(U)$. Among these elements there are 180° rotations and thus the proof of our theorem is finished.

3. *Elliptic spaces*

If the reader scrutinizes the preceding proof for $n = 3$ he will notice a very peculiar feature: In concluding that "small" displacements on the unit sphere can be combined to give arbitrary displacements we have used the archimedean axiom in the field of real numbers. The following example shows that the use of the archimedean axiom is essential.

Let k be an ordered field which contains infinitely small elements (elements > 0 which are smaller than any positive rational number). The field of power series in t with rational coefficients is an example of such a field if we order it by the "lowest term" (see Chapter I, §9). Take again $n = 3$, and base the geometry of V on the quadratic form $x^2 + y^2 + z^2$. It is intuitively clear that the "infinitesimal rotations" will form an invariant subgroup of O_3^+ . Hence O_3^+ will not be simple. We devote this paragraph to a generalization of this example.

Suppose that a valuation of k is given whose value group will be denoted by S (an ordered group with zero-element). We assume S to be non-trivial (which means S does not only consist of 0 and 1) and to be the full image of k under the valuation.

Let V be a space over k with a non-singular orthogonal geometry and assume $n = \dim V \geq 3$.

Let $A_1, A_2, \cdots, A_n$ be a basis of V, $g_{ij} = A_iA_j$ and (h_{ij}) the inverse of the matrix (g_{ij}). Let

$$X = \sum_{\nu=1}^{n} A_\nu x_\nu , \qquad Y = \sum_{\nu=1}^{n} A_\nu y_\nu$$

be vectors of V. We have

$$XA_i = \sum_{\nu=1}^{n} g_{\nu i} x_\nu$$

hence

$$x_i = \sum_{\nu=1}^{n} h_{\nu i}(XA_\nu). \tag{5.9}$$

We define a "norm" $\|X\|$ of X with respect to our valuation by

$$\|X\| = \operatorname{Max}_i |x_i| \tag{5.10}$$

and get from (5.9) (if $s_1 \in S$ is the maximum of $|h_{ij}|$)

$$\|X\| \leq s_1 \cdot \operatorname{Max}_\nu |XA_\nu|. \tag{5.11}$$

If $s_2 = \operatorname{Max}_{i,j} |g_{ij}|$, then $XY = \sum_{i,j=1}^{n} g_{ij}x_iy_j$ shows

$$|XY| \leq s_2 \, \|X\| \cdot \|Y\|. \tag{5.12}$$

Definition 5.2. V is called elliptic with respect to our valuation, if there exists an $s_3 \in S$ such that

$$|XY|^2 \leq s_3 \, |X^2| \cdot |Y^2| \tag{5.13}$$

for all $X, Y \in V$.

Remark. Suppose that V contains an isotropic vector $N \neq 0$. Let N, M be a hyperbolic pair. Then (5.13) can not hold for $X = N$, $Y = M$ and we see that V is *not* elliptic. If V does not contain isotropic vectors $\neq 0$, then it may or may not be elliptic with respect to our valuation.

LEMMA 5.1. *V is elliptic if and only if there exists an $s_4 \varepsilon S$ such that $|X^2| \leq 1$ implies $\|X\| < s_4$.*

Proof:

1) Suppose V is elliptic and $|X^2| \leq 1$. If we put $Y = A_\nu$ in (5.13) we obtain that $|XA_\nu|^2 \leq s_3 |A_\nu^2|$, hence, from (5.11), that $\|X\|^2 \leq s_1^2 s_3 \operatorname{Max}_\nu |A_\nu^2| < s_4^2$ for a suitably selected s_4.

2) Suppose $|X^2| \leq 1$ implies $\|X\| < s_4$. Then $\|Y\| = s_4$ would certainly imply $|Y^2| > 1$. Let X be any non-zero vector. Find $a \varepsilon k$ such that $\|X\| = |a|$ and $b \varepsilon k$ such that $s_4 = |b|$. The vector $(b/a)X$ has norm s_4 and, consequently, $|((b/a)X)^2| > 1$ or $s_4^2 |X^2| > \|X\|^2$. For all $X \varepsilon V$ we have, therefore, $\|X\|^2 \leq s_4^2 |X^2|$. From (5.12) we get now

$$|XY|^2 \leq s_2^2 s_4^4 |X^2| \cdot |Y^2|,$$

which implies that V is elliptic.

LEMMA 5.2. *Suppose that V contains a vector $A \neq 0$ such that there exists an $s_5 \varepsilon S$ satisfying*

$$|XA|^2 \leq s_5 |X^2|.$$

Then V is elliptic.

Proof: If σ ranges over all isometries, then the vectors σA will generate a subspace of V which is invariant under all isometries. By Theorem 3.24 this subspace must be V itself. We may, therefore, select a basis $A_i = \sigma_i A$ of V from this set. Then

$$|XA_i|^2 = |\sigma_i^{-1}X \cdot \sigma_i^{-1}A_i|^2 = |\sigma_i^{-1}X \cdot A|^2 \leq s_5 |(\sigma_i^{-1}X)^2| = s_5 |X^2|$$

and from (5.11)

$$\|X\|^2 \leq s_1^2 s_5 |X^2|.$$

Thus if $|X^2| \leq 1$, then $\|X\|^2 \leq s_1^2 s_5 < s_4^2$ for a suitably selected s_4. V is elliptic.

The following definition still works for any space.

DEFINITION 5.3. An isometry σ of V is called infinitesimal of order s ($s \neq 0$, $s \varepsilon S$) if

$$|X(\sigma Y - Y)|^2 \leq s |X^2| \cdot |Y^2| \quad \text{for all } X, Y \varepsilon V.$$

We shall denote by G_s the set of all infinitesimal isometries of a given order s. Clearly identity lies in G_s .

THEOREM 5.4. *G_s is an invariant subgroup of $O(V)$.*

Proof:

1) Let $\sigma, \tau \in G_s$, then

$$|X(\sigma\tau Y - Y)| = |X(\sigma\tau Y - \tau Y) + X(\tau Y - Y)|$$
$$\leq \text{Max}(|X(\sigma(\tau Y) - \tau Y)|, |X(\tau Y - Y)|),$$
$$|X(\sigma\tau Y - Y)|^2 \leq \text{Max}(s|X^2|\cdot|(\tau Y)^2|, s\,|X^2|\cdot|Y^2|)$$
$$= s\,|X^2|\cdot|Y^2|$$

which shows $\sigma\tau \in G_s$.

2) Let $\sigma \in G_s$, then

$$|X(\sigma^{-1}Y - Y)|^2 = |\sigma X(Y - \sigma Y)|^2 \leq s\,|(\sigma X)^2|\cdot|Y^2|$$
$$= s\,|X^2|\cdot|Y^2|$$

so that $\sigma^{-1} \in G_s$.

3) Let $\sigma \in G_s$, $\tau \in O(V)$, then

$$|X(\tau\sigma\tau^{-1}Y - Y)|^2 = |\tau^{-1}X(\sigma\tau^{-1}Y - \tau^{-1}Y)|^2 \leq s\,|(\tau^{-1}X)^2||(\tau^{-1}Y)^2|$$
$$= s\,|X^2|\cdot|Y^2|$$

which proves that G_s is an invariant subgroup of $O(V)$.

THEOREM 5.5. *As s decreases the sets G_s decrease and we have*

$$\bigcap_s G_s = 1.$$

Proof: The first statement is obvious. To prove the last it suffices to exhibit a G_s which does not contain a given $\sigma \neq 1$.

Select a Y such that $\sigma Y - Y \neq 0$ and an X such that $X(\sigma Y - Y) \neq 0$. Then

$$|X(\sigma Y - Y)|^2 \leq s\,|X^2|\cdot|Y^2|$$

will be false for a small enough $s \in S$.

THEOREM 5.6. *If V is not elliptic, then $G_s = 1$ for any $s \in S$.*

Proof: It suffices to show that if $\sigma \neq 1$ is contained in some G_s , then V is elliptic.

Select Y so that $A = \sigma Y - Y \neq 0$. Then

$$|XA|^2 \leq s\, |X^2| \cdot |Y^2| = s_5\, |X^2|$$

for all $X \varepsilon V$ and a suitable s_5. Lemma 5.2 shows that V is elliptic.

We may, therefore, assume from now on that V is an elliptic space.

THEOREM 5.7. *If V is elliptic, then every G_ε contains elements $\sigma \neq 1$.*

Proof: The idea is the following. Denote by τ_A the symmetry with respect to $\langle A \rangle^*$ and let A and B be vectors which are independent but "very close"; then $\tau_A \neq \tau_B$, hence $\sigma = \tau_B\tau_A \neq 1$ and σ should be "nearly" identity. For a proof we would have to estimate

$$X(\sigma Y - Y) = X(\tau_B\tau_A Y - Y) = \tau_B X(\tau_A Y - \tau_B Y)$$

in terms of $|Y^2|$ and $|X^2| = |(\tau_B X)^2|$. It is, therefore, enough to estimate $X(\tau_A Y - \tau_B Y)$ in terms of $|X^2|$ and $|Y^2|$.

For $\tau_A Y$ we have a formula:

$$\tau_A Y = Y - \frac{2(AY)}{A^2} \cdot A.$$

One sees this by observing that the right side is a linear map which is identity on $\langle A \rangle^*$ and sends A into $-A$. Therefore

$$\tau_A Y - \tau_B Y = \frac{2(BY)}{B^2} \cdot B - \frac{2(AY)}{A^2} \cdot A$$

and

$$X(\tau_A Y - \tau_B Y) = \frac{2}{A^2B^2}\left(A^2(BX)(BY) - B^2(AX)(AY)\right). \tag{5.14}$$

We select a fixed $A \neq 0$ and a fixed C independent of A. Choose $\eta \neq 0$ in k and set $B = A + \eta C$ which will achieve $\tau_A \neq \tau_B$. This η will be chosen "very small".

We have $B^2 = A^2 + 2\eta(AC) + \eta^2C^2$ and if $|\eta|$ is small enough $|A^2|$ is the maximal term; from $A^2 = B^2 - 2\eta(AC) - \eta^2C^2$ one sees that for small $|\eta|$ the term $|B^2|$ must be maximal, since $A^2 \neq 0$. This shows $|A^2| = |B^2|$ so that the factor $2/A^2B^2$ of (5.14) is merely a constant in absolute value and may be disregarded.

If $B = A + \eta C$ is substituted in the parentheses on the right side of (5.14) the term $A^2(AX)(AY)$ will cancel and a polynomial in η will remain which has a factor η. Factoring out this η, the coefficients of the remaining polynomial in η will be, apart from constant factors, products of either (AX) or (CX) by either (AY) or (CY). This shows that we can find an $s' \varepsilon S$ such that for small $|\eta|$ we have

$$|X(\tau_A Y - \tau_B Y)| \leq s' \cdot |\eta| \cdot \mathrm{Max}(|AX|, |CX|) \cdot \mathrm{Max}(|AY|, |CY|).$$

Since the space is elliptic, we obtain

$$|AX|^2 \leq s_3 \, |A^2| \cdot |X^2|, \qquad |CX|^2 \leq s_3 \, |C^2| \cdot |X^2|$$

and see that an $s'' \varepsilon S$ can be found such that

$$|X(\tau_A Y - \tau_B Y)| \leq s'' \, |\eta| \cdot |X^2| \cdot |Y^2|.$$

All we have to do is to select $|\eta|$ small enough.

THEOREM 5.8. *If $|s| < |4|$, then G_s does not contain isometries σ which reverse some non-zero vector. It follows that then $G_s \subset O^+(V)$, that G_s does not contain any element of even order and (if V is elliptic) that $G_s \cap \Omega(V) \neq 1$.*

Proof: Suppose $\sigma A = -A$, $A \neq 0$ and V elliptic. Put $X = Y = A$. Then

$$|X(\sigma Y - Y)|^2 = |4| \cdot |A^2|^2 = |4| \cdot |X^2| \cdot |Y^2|$$

which shows the first part of the contention.

If σ is a reflexion and $n = \dim V$ is odd, then $-\sigma$ is a rotation. Since n is odd, $-\sigma$ keeps some non-zero vector fixed, hence σ will reverse it. If n is even and σ is a reflexion, σ is a product of at most $n - 1$ symmetries and keeps, therefore, a vector $B \neq 0$ fixed. V is elliptic, $B^2 \neq 0$. The reflection σ has the form $1_{\langle B \rangle} \perp \tau$ where τ is a reflexion on the odd-dimensional space $\langle B \rangle^*$. It follows again that σ reverses some non-zero vector.

If σ has order $2r$, then σ^r is an involution $\neq 1$ and will reverse some vector; σ^r can not be in G_s, hence σ can not be in it either. It follows that if $\sigma \varepsilon G_s$ and $\sigma \neq 1$, then $\sigma^2 \varepsilon G_s$ and $\sigma^2 \neq 1$. But $\sigma^2 \varepsilon \Omega(V)$.

The groups $G_s \cap \Omega(V)$ form now a descending chain of invariant non-trivial subgroups of $\Omega(V)$. Therefore, if V is elliptic, $\Omega(V)$ is not simple either.

THEOREM 5.9. *If $\sigma \varepsilon G_s$, $\tau \varepsilon G_t$, then the commutator $\sigma\tau\sigma^{-1}\tau^{-1} \varepsilon G_{st}$.*

Proof: We have $|X(\sigma - 1)Y|^2 \leq s \, |X^2| \cdot |Y^2|$. Put $X = (\sigma - 1)Y$. We obtain $|((\sigma - 1)Y)^2| \leq s \, |Y^2|$ and consequently

$$|X(\sigma - 1)(\tau - 1)Y|^2 \leq s \, |X^2| \cdot |((\tau - 1)Y)^2| \leq st \, |X^2| \cdot |Y^2|$$

and

$$|X\cdot(\tau - 1)(\sigma - 1)Y|^2 \leq st\ |X^2|\cdot|Y^2|.$$

Since $\sigma\tau - \tau\sigma = (\sigma - 1)(\tau - 1) - (\tau - 1)(\sigma - 1)$ we obtain

$$|X\cdot(\sigma\tau - \tau\sigma)Y|^2 \leq st\ |X^2|\cdot|Y^2|.$$

If we replace Y by $\sigma^{-1}\tau^{-1}Y$, we get

$$|X\cdot(\sigma\tau\sigma^{-1}\tau^{-1} - 1)Y|^2 \leq st\ |X^2|\cdot|Y^2|$$

and this proves our contention.

This theorem is significant only if $s < 1$, $t < 1$. We mention as a special case that for $s < 1$ the commutator of two elements of G_s lies in G_{s^2} so that the factor group G_s/G_{s^2} is abelian.

THEOREM 5.10. *If V is elliptic and s is sufficiently large, then* $G_s = O(V)$.

Proof: Since we assume V elliptic we have

$$|XY|^2 \leq s_3\ |X^2|\cdot|Y^2| \quad \text{and} \quad |X\cdot\sigma Y|^2 \leq s_3\ |X^2|\cdot|Y^2|,$$

hence $|X(\sigma Y - Y)|^2 \leq s_3\ |X^2|\cdot|Y^2|$ which shows that any $\sigma \in G_{s_3}$.

It would be of interest to know the structure of factor groups also for $s > 1$.

To give some examples let us turn to the one we gave at the beginning of this section. If k is an ordered field which is non-archimedean we can associate with the ordering a valuation as described in Chapter I, §10. The finite elements of k are those which satisfy $|a| \leq 1$.

We had based the geometry on the quadratic form $x^2 + y^2 + z^2$. If $|x^2 + y^2 + z^2| \leq 1$, then $x^2 + y^2 + z^2 \leq m$ where m is an integer. Hence $0 \leq x^2 \leq x^2 + y^2 + z^2 \leq m$. It follows that $|x^2| \leq 1$ and, consequently, $|x| \leq 1$. This shows that $|X^2| \leq 1$ implies $\|X\| \leq 1$. Our geometry is elliptic.

To give a second example let k be the field Q of rational numbers and base the geometry again on the quadratic form $x^2 + y^2 + z^2$. Select as valuation of Q the 2-adic one, where $|a| \leq 1$ means that a has an odd denominator. Let d be the least common denominator of x, y, z and $x = r/d$, $y = s/d$, $z = t/d$. Suppose $|x^2 + y^2 + z^2| = |(r^2 + s^2 + t^2)/d^2| \leq 1$. This implies that $(r^2 + s^2 + t^2)/d^2$ can also be written with an odd denominator. We contend that d itself must be odd. Indeed, if d were even, then $r^2 + s^2 + t^2$ would have to be devisible by 4. One sees easily that this is only possible if r, s and t are even. But then d would not be the *least* common denominator of x, y, z. Thus we have shown that $|x| \leq 1$, $|y| \leq 1$, $|z| \leq 1$ and $|X^2| \leq 1$ implies again that $\|X\| \leq 1$, our geometry is elliptic.

4. *The Clifford algebra*

Let V be a vector space with a non-singular orthogonal geometry. For deeper results on orthogonal groups we shall have to construct a certain associative ring called the Clifford algebra $C(V)$ of V.

We must first introduce some elementary notions of set theory. If $S_1, S_2, \cdots, S_r$ are subsets of a given set M we define a sum $S_1 + S_2 + \cdots + S_r$ which will *not* be the union. It shall consist of those elements of M which occur in an *odd* number of the sets S_i. The following rules are easily derived:

$$(S_1 + S_2 + \cdots + S_r) + S_{r+1} = S_1 + S_2 + \cdots + S_{r+1},$$

$$(S_1 + S_2 + \cdots + S_r) \cap T = (S_1 \cap T) + (S_2 \cap T) + \cdots + (S_r \cap T).$$

The empty set shall be denoted by $\emptyset$.

Let now $A_1, A_2, \cdots, A_n$ be an orthogonal basis of V. The sets in question are the 2^n subsets of the set $M = \{1, 2, \cdots, n\}$. We construct a vector space $C(V)$ over k of dimension 2^n with basis elements e_S, one e_S for each subset S of M.

In $C(V)$ a multiplication denoted by $\circ$ is defined between the basis elements e_S and extended to all of $C(V)$ by linearity. The definition is:

$$e_S \circ e_T = \prod_{\substack{s \in S \\ t \in T}} (s, t) \cdot \prod_{i \in S \cap T} A_i^2 \cdot e_{S+T}.$$

The symbol (s, t) is a sign; it shall be $+1$ if $s \leq t$ and -1 if $s > t$. The term A_i^2 is the ordinary square of the basis vector A_i of V and is, therefore, an element of k. If S, T or $S \cap T$ are empty, then "empty products" occur in the definition; they have to be interpreted as usual, namely as 1.

This definition will become understandable in a short while.

The main problem is the associativity. We have

$$(e_S \circ e_T) \circ e_R = \prod_{\substack{s \in S \\ t \in T}} (s, t) \cdot \prod_{\substack{j \in S+T \\ r \in R}} (j, r) \cdot \prod_{i \in S \cap T} A_i^2 \cdot \prod_{\lambda \in (S+T) \cap R} A_\lambda^2 \cdot e_{S+T+R}$$

and we shall rewrite the right side in a more symmetric form.

First the signs: If we let j range not over $S + T$ but first over all of S and then over all of T, then any $j \in S \cap T$ will appear twice. No change is produced in this way since $(j, r)^2 = 1$. The signs can, therefore, be written as

$$\prod_{\substack{s \in S \\ t \in T}} (s, t) \cdot \prod_{\substack{s \in S \\ r \in R}} (s, r) \cdot \prod_{\substack{t \in T \\ r \in R}} (t, r),$$

which is a more satisfactory form.

Now the A_ν^2: We have $(S + T) \cap R = (S \cap R) + (T \cap R)$. If ν belongs to all three sets S, T, R, then it is in $S \cap T$ but not in $(S \cap R) + (T \cap R)$. If ν belongs to S and T but not to R, then it is in $S \cap T$ but not in $(S \cap R) + (T \cap R)$. If ν lies in S and R but not in T, or in T and R but not in S, then it is not in $S \cap T$ but in $(S \cap R) + (T \cap R)$. If, finally, ν is in only one of the sets S, T, R or in none of them, then ν will neither be in $S \cap T$ nor in $(S \cap R) + (T \cap R)$. We must, therefore, take the product over those A_ν^2 for which ν appears in *more* than one of the sets S, T, R.

The form we have achieved is so symmetric that $(e_S \circ e_T) \circ e_R = e_S \circ (e_T \circ e_R)$ is rather obvious. Our multiplication makes $C(V)$ into an associative ring, the Clifford algebra of V. One sees immediately that $e_\phi \circ e_S = e_S \circ e_\phi = e_S$ which shows that $C(V)$ has the unit element e_ϕ which we also denote by 1. The scalar multiples of e_ϕ : $k \cdot e_\phi$ shall be identified with k.

We shall also identify V with a certain subspace of $C(V)$ as follows: the vector A_i is identified with the vector $e_{\{i\}}$ (associated with the set $\{i\}$ containing the single element i). The vector $X = \sum_{i=1}^n x_i A_i$ of V is identified with the vector $\sum_{i=1}^n x_i e_{\{i\}}$ of $C(V)$. We must now distinguish the old scalar product XY in V and the product $X \circ Y$ in $C(V)$. We find

$$A_i \circ A_i = e_{\{i\}} \circ e_{\{i\}} = (i, i) A_i^2 e_\phi = A_i^2 .$$

If $i \neq j$, then

$$(A_i \circ A_j) + (A_j \circ A_i) = (i, j) e_{\{i,j\}} + (j, i) e_{\{i,j\}} = 0.$$

Let $X = \sum_{i=1}^{n} x_i A_i$, $Y = \sum_{j=1}^{n} y_j A_j$; then

$$(X \circ Y) + (Y \circ X) = \sum_{i,j=1}^{n} x_i y_j ((A_i \circ A_j) + (A_j \circ A_i))$$

$$= 2 \sum_{i=1}^{n} x_i y_i A_i^2 = 2(XY),$$

where (XY) is the old scalar product in V. The rule

$$(5.14) \qquad (X \circ Y) + (Y \circ X) = 2(XY)$$

has two special cases. The commutation rule

$$(5.15) \qquad (X \circ Y) = -(Y \circ X)$$

if X and Y are orthogonal, and

$$(5.16) \qquad X \circ X = X^2.$$

Let $S \subset M$ be non-empty. Arrange the elements of S in an increasing sequence: $i_1 < i_2 < \cdots < i_r$. Then

$$e_S = A_{i_1} \circ A_{i_2} \circ \cdots \circ A_{i_r} ,$$

as one shows easily by induction. The vectors A_i are, therefore, generators of the Clifford algebra. We also understand now the definition of the multiplication $e_S \circ e_T$: if one writes e_S and e_T as products of the basis vectors A_i , one has to use the commutation rule a certain number of times and then possibly the rule (5.16) in order to express $e_S \circ e_T$ as multiple of e_{S+T} . The result would be our definition[1].

$C(V)$ contains two subspaces $C^+(V)$ and $C^-(V)$ of dimension 2^{n-1}. $C^+(V)$ is spanned by the vectors e_S for which S contains an even number of elements, $C^-(V)$ by those e_S for which S contains an odd number of elements. If both S and T contain an even number or if both contain an odd number of elements, then $S + T$ contains an even number. If one of them contains an even and the other an odd number of elements, then $S + T$ will contain an odd number of elements. This leads to the rules:

$$C^+(V) \circ C^+(V) \subset C^+(V), \qquad C^-(V) \circ C^-(V) \subset C^+(V),$$

$$C^-(V) \circ C^+(V) \subset C^-(V), \qquad C^+(V) \circ C^-(V) \subset C^-(V).$$

[1] It is now also easy to see that $C(V)$ does not depend on the choice of the orthogonal basis of V.

Obviously, $C(V) = C^+(V) \oplus C^-(V)$. Since $1 = e_\phi$ belongs to $C^+(V)$ this subspace is a subalgebra of $C(V)$ with a unit element. It is clear that $C^+(V)$ is generated by those e_T for which T contains two elements.

Each basis vector e_S has an inverse which is just a scalar multiple of it, as is obvious from

$$e_S \circ e_S = \prod_{s_1, s_2 \varepsilon S} (s_1, s_2) \cdot \prod_{i \varepsilon S} A_i^2 \ \varepsilon\ k^*.$$

The product $e_S \circ e_T$ differs only in sign from $e_T \circ e_S$. If we denote by $[S]$ the number of elements in S, then we find for the sign

$$\prod (s, t)(t, s) = (-1)^{[S][T]-[S\cap T]}.$$

We may write this result in the form

$$e_T \circ e_S \circ e_T^{-1} = (-1)^{[S][T]-[S\cap T]} \cdot e_S .$$

We shall study the sign change which e_T produces more closely. Let e_S be a given basis element and T range over all sets with two elements, $[T] = 2$; is there a T which produces a sign change? We find as condition that $[S \cap T] = 1$, i.e., that one element of T should be in S and the other not. Such a T will exist unless S is either $\emptyset$ or the whole set M.

An element $\alpha \ \varepsilon\ C(V)$ can be written in the form $\alpha = \sum_S \gamma_S e_S$ with $\gamma_S \ \varepsilon\ k$. We wish to find those α which are in the centralizer of $C^+(V)$, and which, therefore, commute with *all* e_T where $[T] = 2$. Then $e_T \circ \alpha \circ e_T^{-1} = \alpha$ or

$$\sum_S \gamma_S(e_T \circ e_S \circ e_T^{-1}) = \sum_S \gamma_S e_S .$$

For a given $S \neq \emptyset, M$, one can find an e_T which produces a sign change, hence $\gamma_S = 0$ for these S. The centralizer of $C^+(V)$ is, therefore,

$$C_0(V) = k + k e_M .$$

$C_0(V)$ is a subalgebra of $C(V)$ with a very simple structure. We find

$$e_M \circ e_M = (-1)^{n(n-1)/2} A_1^2 A_2^2 \cdots A_n^2 = (-1)^{n(n-1)/2} G$$

where G is the discriminant of V.

If n is odd, then e_M is not in $C^+(V)$ and the center of $C^+(V)$ reduces to k. If n is even, then $C_0(V) \subset C^+(V)$, $C_0(V)$ is the center of $C^+(V)$.

To find the center of $C(V)$ we must see whether e_M commutes with all e_T where $[T] = 1$. One finds that this is the case if and only if n is odd. The center of $C(V)$ is, therefore, k if n is even and $C_0(V)$ if n is odd.

The question which elements of $C^+(V)$ commute with all elements of V has a uniform answer: they are the elements of k in all cases. Indeed, they must commute with every element of $C(V)$ and must lie in $C^+(V)$.

We shall also have to find out which elements $\alpha = \sum_S \gamma_S e_S$ of $C(V)$ anticommute with *every* vector of V, i.e., satisfy $\alpha \circ A_i = -(A_i \circ \alpha)$ for all i. If $\gamma_S \neq 0$, then $e_T \circ e_S \circ e_T^{-1}$ must equal $-e_S$ for all T with $[T] = 1$. This gives as condition that $[S] - [S \cap T]$ must be odd for all such T. Only $S = M$ could possibly do this and then only if n is even. The answer to our question is, therefore, $\alpha = 0$ if n is odd and $\alpha \in ke_M$ if n is even.

If α is an element of $C(V)$ which has an inverse, then one abbreviates $\alpha \circ \xi \circ \alpha^{-1}$ by the symbol ξ^α. The map $C(V) \to C(V)$ given by $\xi \to \xi^\alpha$ is an automorphism of $C(V)$ which satisfies $(\xi^\beta)^\alpha = \xi^{\alpha\beta}$. We denote by $R(V)$ the set of all α which map the subspace V of $C(V)$ into itself, i.e., for which $X \in V$ implies $X^\alpha \in V$. Let α be such an element and let us apply the map to both sides of equation (5.14). The right side, $2(XY)$, remains fixed, being an element of k; the left side becomes $(X^\alpha \circ Y^\alpha) + (Y^\alpha \circ X^\alpha)$ which is $2(X^\alpha Y^\alpha)$. We obtain $(XY) = (X^\alpha Y^\alpha)$ which shows that α induces an isometry on V, i.e., an element σ_α of the group $O(V)$.

The map $\alpha \to \sigma_\alpha$ is a homomorphism of the group $R(V)$ into $O(V)$. The kernel of this map consists of those elements α of the center of $C(V)$ which have an inverse. If n is even, this kernel is k^*; if n is odd, it is the set $C_0^*(V)$ of all elements of $C_0(V)$ which have an inverse. We shall now determine the image of our map.

Let $B, X \in V$ and $B^2 \neq 0$. Equation (5.14) shows $(B \circ X) = 2(BX) - (X \circ B)$. Equation (5.16) implies that $B^{-1} = (1/B^2)\cdot B$; if we multiply the preceding equation on the right by B^{-1}, then the left side becomes X^B and we obtain

$$X^B = \frac{2(BX)}{B^2} B - X = -\tau_B(X)$$

where $\tau_B(X) = X - (2(BX)/B^2)\, B$. The map $\tau_B : V \to V$ is identity on the hyperplane $\langle B \rangle^*$ and sends the vector B into $-B$. It is, therefore, the symmetry with respect to $\langle B \rangle^*$.

We remember that every isometry σ of V is a product $\sigma = \tau_{B_1}\tau_{B_2} \cdots \tau_{B_r}$ of symmetries. The fact that one can achieve $r \leq n$ will not be used right now. If $\sigma \varepsilon O^+(V)$, then r will be even, otherwise it is odd. We have found

$$X^B = -\tau_B(X)$$

and obtain, by raising X to the powers $B_r, B_{r-1}, \cdots, B_1$ successively, that

$$X^{B_1 \circ B_2 \circ \cdots \circ B_r} = (-1)^r \sigma(X).$$

If n is even, then -1_V is a rotation and we see that $(-1)^r\sigma$ can range over all of $O(V)$. If σ is a rotation, then $\sigma(X)$ is expressed directly; if σ is a reflexion, the reflexion $-\sigma$ is given by our formula. The map $R(V) \to O(V)$ is onto, the kernel of this map was k^* and we see that every element of $R(V)$ has the form $\gamma \cdot B_1 \circ B_2 \circ \cdots \circ B_r$; since $\gamma \varepsilon k^*$ can be united with B_1 we may say that the elements of $R(V)$ are merely products $B_1 \circ B_2 \circ \cdots \circ B_r$ of non-isotropic vectors. If r is even, then the element lies in $C^+(V)$, otherwise it is in $C^-(V)$. Rotations are only produced by the elements of $C^+(V)$, reflexions by those of $C^-(V)$.

If n is odd, then -1_V is a reflexion and the map $(-1)^r\sigma$ is always a rotation. This raises the question whether one can find an $\alpha \varepsilon R(V)$ such that $X^\alpha = \sigma(X)$ is a reflexion. Setting $\beta = B_1 \circ B_2 \circ \cdots \circ B_r$ (where r is odd) the element $\beta^{-1} \circ \alpha = \gamma$ of $R(V)$ would induce the map $X^\gamma = -X$. But we have seen that no $\gamma \neq 0$ can anticommute with every element of V if n is odd. The result is that the image of $R(V)$ consists only of rotations. These rotations can already be obtained from elements $B_1 \circ B_2 \circ \cdots \circ B_r$ which will lie in $C^+(V)$ if r is even and in $C^-(V)$ if r is odd. This suggests the

Definition 5.4. An element α of either $C^+(V)$ or of $C^-(V)$ is called regular if α has an inverse and if α maps V onto $V : X \varepsilon V$ implies $X^\alpha \varepsilon V$.

If n is odd, then α and β induce the same rotation if and only if they differ by a factor of $C_0^*(V)$. Should both of them be either in $C^+(V)$ or both in $C^-(V)$, then this factor must be in k^*, since $e_M \varepsilon C^-(V)$. The regular elements are products of non-isotropic vectors in this case also.

We are mainly interested in the group $O^+(V)$ and can restrict ourselves to the regular elements of $C^+(V)$. They form a group which shall be denoted by $D(V)$. We obtain a uniform statement:

THEOREM 5.11. *The regular elements of $C^+(V)$ are products of an even number of non-isotropic vectors of V. The map $D(V) \to O^+(V)$ given by $\alpha \to \sigma_\alpha$ is onto and its kernel is k^*. We have, therefore, an isomorphism: $O^+(V) \simeq D(V)/k^*$.*

We return to the expression

$$e_S = A_{i_1} \circ A_{i_2} \circ \cdots \circ A_{i_r}, \qquad i_1 < i_2 < \cdots < i_r,$$

$$S = \{i_1, i_2, \cdots, i_r\}$$

for non-empty subsets S of M. Let us reverse all the factors and call the resulting product e'_S:

$$e'_S = A_{i_r} \circ A_{i_{r-1}} \circ \cdots \circ A_{i_1}.$$

The commutation rule shows that e'_S differs from e_S only by a sign,

$$e'_S = (-1)^{r(r-1)/2} e_S.$$

In agreement with this rule we set $1' = e'_\phi = e_\phi = 1$ and extend the map to $C(V)$ by linearity. If $\alpha = \sum_S \gamma_S e_S$, then $\alpha' = \sum_S \gamma_S e'_S$.

Let us compare (for any A_i) the products

$$A_i \circ e_S = A_i \circ (A_{i_1} \circ A_{i_2} \circ \cdots \circ A_{i_r})$$

and

$$e'_S \circ A_i = (A_{i_r} \circ A_{i_{r-1}} \circ \cdots \circ A_{i_1}) \circ A_i.$$

In order to express the first product as a multiple of an e_T one has to use the commutation rule a certain number of times to bring A_i into the correct position and then possibly the rule $A_i \circ A_i = A_i^2$, if i occurs among the i_ν.

Expressing the second product as a multiple of an e'_T we see that we have to use the same number of interchanges followed by a possible replacement of $A_i \circ A_i$ by A_i^2. This shows:

Should $A_i \circ e_S = \gamma e_T$, then $e'_S \circ A_i = \gamma e'_T$. This is the special case of the rule

$$e'_S \circ e'_T = (e_T \circ e_S)'$$

if T contains only one element. By induction one can now prove this rule for any T.

The map $J : C(V) \to C(V)$ given by $\alpha^J = \alpha'$ is, therefore, an antiautomorphism of $C(V)$.

One is lead to introduce a norm on $C(V)$:

$$N(\alpha) = \alpha \circ \alpha^J.$$

This norm does *not* satisfy in general the rule $N(\alpha \circ \beta) = N(\alpha) \circ N(\beta)$. It does so, however, if α and β are regular. Indeed, if

$$\alpha = B_1 \circ B_2 \circ \cdots \circ B_r ,$$

then

$$N(\alpha) = \alpha \circ \alpha^J = (B_1 \circ B_2 \circ \cdots \circ B_r) \circ (B_r \circ B_{r-1} \circ \cdots \circ B_1)$$

and from $B_r \circ B_r = B_r^2$, $B_{r-1} \circ B_{r-1} = B_{r-1}^2$, $\cdots$ one obtains

$$(5.17) \qquad N(\alpha) = B_1^2 B_2^2 \cdots B_r^2 \in k^*.$$

This implies $N(\alpha \circ \beta) = \alpha \circ \beta \circ \beta^J \circ \alpha^J = N(\beta) \cdot (\alpha \circ \alpha^J) = N(\alpha)N(\beta)$ if α and β are regular.

We are now in a position to prove the main lemma:

LEMMA 5.3. *Let B_1 , B_2 , $\cdots$, B_r be non-isotropic vectors of V, τ_{B_i} the summetry with respect to $\langle B_i \rangle^*$ and assume $\tau_{B_1} \cdot \tau_{B_2} \cdot \cdots \cdot \tau_{B_r} = 1_V$. The product $B_1^2 B_2^2 \cdots B_r^2$ is then the square of an element of k^*.*

Proof: Since 1_V is a rotation, we know that r is even. The element $B_1 \circ B_2 \circ \cdots \circ B_r$ of $C^+(V)$ induces the identity on V and is, therefore, in the kernel k^* of the map $D(V) \to O^+(V)$. If we set $a = B_1 \circ B_2 \circ \cdots \circ B_r$, then $a \in k^*$. The norm of a is $a \circ a^J = a \cdot a = a^2$ on one hand and, by (5.17), $B_1^2 B_2^2 \cdots B_r^2$ on the other.

5. *The spinorial norm*

Let $\sigma = \tau_{A_1} \tau_{A_2} \cdots \tau_{A_r} = \tau_{B_1} \tau_{B_2} \cdots \tau_{B_s}$ be two expressions for an isometry σ as product of symmetries. Then

$$\tau_{A_1} \tau_{A_2} \cdots \tau_{A_r} \tau_{B_s} \tau_{B_{s-1}} \cdots \tau_{B_1} = 1_B$$

and Lemma 5.3 implies that $A_1^2 A_2^2 \cdots A_r^2 B_s^2 B_{s-1}^2 \cdots B_1^2$ is a square of k^*. The product $A_1^2 A_2^2 \cdots A_r^2$ differs, therefore, from $B_1^2 B_2^2 \cdots B_s^2$ by the square factor

$$\frac{A_1^2 A_2^2 \cdots A_r^2 B_1^2 B_2^2 \cdots B_s^2}{(B_1^2 B_2^2 \cdots B_s^2)^2} .$$

Let us denote by k^{*2} the multiplicative group of squares of k^*. Then we have shown that the map

$$\theta(\sigma) = A_1^2 A_2^2 \cdots A_r^2 \cdot k^{*2} \tag{5.18}$$

is well defined. Since obviously $\theta(\sigma_1\sigma_2) = \theta(\sigma_1)\theta(\sigma_2)$ it gives a homomorphism $O(V) \to k^*/k^{*2}$. Since the image group is commutative, the commutator group $\Omega(V)$ is contained in the kernel.

It is, however, not yet a reasonable map. If we replace the scalar product XY on V by a new product $t \cdot (XY)$, where t is a fixed element of k^*, then the orthogonal group does not change, isometries are still the same linear maps of V. The map $\theta(\sigma)$ does change, however, by the factor t^r. If r is odd, $\theta(\sigma)$ changes radically and only for even r does it stay the same. One should, therefore, restrict our map to rotations. Since $\Omega(V) \subset O^+(V)$ the group $\Omega(V)$ is still contained in the kernel $O'(V)$ of this restricted map.

Definition 5.5. The map (5.18) of $O^+(V)$ into k^*/k^{*2} is called the spinorial norm. The image consists of all cosets of the form $A_1^2 A_2^2 \cdots A_r^2 k^{*2}$ with an even number r of terms. The "spinorial kernel" $O'(V)$ of this map satisfies $\Omega(V) \subset O'(V) \subset O^+(V)$.

Aside from the map θ we introduce the canonical map $f : O(V) \to O(V)/\Omega(V)$ and state the main property of f to be used.

Theorem 5.12. *If* $\sigma = \tau_{A_1}\tau_{A_2} \cdots \tau_{A_r} \in O(V)$, *then* $f(\sigma)$ *depends only on the set* $\{A_1^2, A_2^2, \cdots, A_r^2\}$ *and neither on the particular vectors* $A_1, A_2, \cdots, A_r$ *nor on their order.*

Proof: Suppose $A_i^2 = B_i^2$. There exists a $\lambda_i \in O(V)$ such that $\lambda_i A_i = B_i$. Then $\lambda_i \tau_{A_i} \lambda_i^{-1} = \tau_{B_i}$. Since the image of f is a commutative group we have $f(\tau_{B_i}) = f(\lambda_i \tau_{A_i} \lambda_i^{-1}) = f(\lambda_i) f(\tau_{A_i}) f(\lambda_i)^{-1} = f(\tau_{A_i})$ hence $f(\sigma) = f(\prod_i \tau_{B_i})$ where any arrangement of the product of the τ_{B_i} gives the same result.

Theorem 5.13. *If* $V = U \perp W$, $\sigma = \tau \perp \rho$ *where* $\tau \in O^+(U)$ and $\rho \in O^+(W)$, *then* $\theta(\sigma) = \theta(\tau)\theta(\rho)$. *If we identify* $\tau \in O^+(U)$ *with* $\tau \perp 1_W \in O^+(V)$, *then* $\theta(\tau)$ *computed in* U *is the same as* $\theta(\tau)$ *computed in* V. *We may say that the spinorial norm is invariant under orthogonal imbedding. We have* $O'(U) = O^+(U) \cap O'(V)$.

Proof: Since $\sigma = (\tau \perp 1_W)(1_U \perp \rho)$ it is enough to show $\theta(\tau) = \theta(\tau \perp 1_W)$. Let $\tau = \tau'_{A_1}\tau'_{A_2} \cdots \tau'_{A_r}$ be expressed as product of symmetries τ'_{A_i} of the space U. Then $\tau_{A_i} = \tau'_{A_i} \perp 1_W$ is the symmetry with respect to $\langle A_i \rangle^*$ of V and $\tau \perp 1_W = \tau_{A_1}\tau_{A_2} \cdots \tau_{A_n}$. All contentions are now obvious.

THEOREM 5.14. *Suppose* $\sigma = \tau_A \tau_B$ *is a product of two symmetries only and* $\sigma \in O'(V)$. *Then* $\sigma \in \Omega(V)$.

If $\dim V = 2$, *then* $O'(V) = \Omega(V)$ *and each element of* $\Omega(V)$ *is a square of a rotation. If* $\dim V = 3$, *then we have also* $O'(V) = \Omega(V)$ *and each element of* $\Omega(V)$ *is the square of a rotation with the same axis.*

Proof:

1) If $\sigma = \tau_A \tau_B$, then $\theta(\sigma) = A^2 B^2 k^{*2}$. If $\sigma \in O'(V)$, then $A^2 B^2 = a^2 \in k^{*2}$. Put $B_1 = (A^2/a)B$, then $\tau_{B_1} = \tau_B$ and $B_1^2 = (A^2)^2 B^2/a^2 = A^2$. We may, therefore, assume from the start that $A^2 = B^2$. But then $f(\tau_A \tau_B) = f(\tau_A \tau_A) = 1$ by Theorem 5.12, hence $\sigma \in \Omega(V)$.

2) If $n = \dim V = 2$ or 3, two symmetries suffice to express any $\sigma \in O^+(V)$. Hence $O'(V) = \Omega(V)$. The squares of rotations generate $\Omega(V)$. If $n = 2$, $O^+(V)$ is commutative, every $\sigma \in \Omega(V)$ is a square. If $n = 3$, $\sigma \in O'(V)$, let $\langle A \rangle$ be the axis of σ. Should A be isotropic, then σ is a square of a rotation with axis $\langle A \rangle$ by our previous discussion of such rotations. If $A^2 \neq 0$, then σ may be identified with the rotation which it induces on the plane $\langle A \rangle^*$ and the contention follows from $\theta(\sigma) = 1$ and the case $n = 2$.

THEOREM 5.15. *If* V *is any space,* U *a non-singular subspace of dimension* 2 *or* 3, $\sigma = \tau \perp 1_{U^*} \in O'(V)$, *then* τ *is a square and* $\sigma \in \Omega(V)$.

Proof: Use $\theta(\sigma) = \theta(\tau)$ and the preceding theorem.

THEOREM 5.16. *Let* U *be a non-singular subspace of* V *with the following properties:*

1) $O'(U) = \Omega(U)$.

2) *To every vector* $A \in V$ *with* $A^2 \neq 0$ *we can find a vector* $B \in U$ *with* $A^2 = B^2$.

Then $O'(V) = \Omega(V)$.

Proof: Let $\sigma = \tau_{A_1}\tau_{A_2} \cdots \tau_{A_r} \in O'(V)$. Find $B_i \in U$ such that $B_i^2 = A_i^2$ and put $\sigma_1 = \tau_{B_1}\tau_{B_2} \cdots \tau_{B_r}$. By Theorem 5.12, $f(\sigma) = f(\sigma_1)$ and clearly $\theta(\sigma) = \theta(\sigma_1) = 1$. Each τ_{B_i} leaves U^* element-wise fixed, $\sigma_1 = \sigma_2 \perp 1_{U^*}$ and $\theta(\sigma_1) = \theta(\sigma_2) = 1$. By assumption $\sigma_2 \in \Omega(U)$, σ_2 is a product of squares; it follows that σ_1 is a products of squares, $f(\sigma_1) = 1$, and therefore $f(\sigma) = 1$, $\sigma \in \Omega(V)$.

THEOREM 5.17. *If* V *contains isotropic vectors* $\neq 0$, *then* $O'(V) = \Omega(V)$.

Proof: Let N, M be a hyperbolic pair of V. The subspace $U = \langle N, M\rangle$ satisfies the conditions of Theorem 5.16: condition 1) since $\dim U = 2$ and condition 2) since $(aN + bM)^2 = 2ab$ may take on any value in k.

THEOREM 5.18. *If V contains isotropic vectors, then the factor group $O^+(V)/\Omega(V)$ is canonically isomorphic (by the spinorial map) to k^*/k^{*2}.*

Proof: The ontoness follows from the proof of the preceding theorem, since the squares of vectors of $\langle N, M\rangle$ range over k.

If V is any space, we can always imbed it in a larger space $\bar{V}$ for which $O'(\bar{V}) = \Omega(\bar{V})$. This can be done in many ways, by setting, for instance, $\bar{V} = V \perp P$ where P is a hyperbolic plane. We could also set $\bar{V} = V \perp L$ where L is a suitable line $L = \langle B\rangle$: for some $A \in V$, $A^2 \neq 0$ prescribe $B^2 = -A^2$, then the vector $B + A$ will be isotropic, $O'(\bar{V}) = \Omega(\bar{V})$. If we write for such a space $\bar{V} = V \perp U$, then, by Theorem 5.13,

$$O'(V) = O^+(V) \cap \Omega(\bar{V}).$$

The geometric interpretation of $O'(V)$ is, therefore, the following: it consists of those rotations of V which fall into the commutator group of the "roomier" space $\bar{V}$.

THEOREM 5.19. *$-1_V \in O'(V)$ if and only if V is even-dimensional and the discriminant G of V is a square. Should V contain isotropic vectors, then these are the conditions for -1_V to be in $\Omega(V)$.*

Proof: If $\dim V$ is odd, then -1_V is a reflexion. Let $\dim V$ be even and $A_1, A_2, \cdots, A_n$ an orthogonal basis of V. Then $-1_V = \tau_{A_1}\tau_{A_2} \cdots \tau_{A_n}$, hence $\theta(-1_V) = A_1^2 A_2^2 \cdots A_n^2 \cdot k^{*2} = Gk^{*2}$.

Up to here everything went fine. But if V does not contain isotropic vectors (is anisotropic), and if $\dim V \geq 4$, then the nature of $\Omega(V)$ is still unknown and even no conjectures for a precise description of $\Omega(V)$ are available. Dieudonné has constructed counterexamples for $n = 4$.

6. *The cases $\dim V \leq 4$*

In these cases the structure of the Clifford algebra is much simpler than in the higher cases. We had seen that $e_S^J = e_S' = (-1)^{r(r-1)/2} e_S$ where r is the number of elements of S. If $n \leq 4$, then $r \leq 4$; e_S is kept fixed by the anti-automorphism J if and only if $r = 0, 1, 4$.

The elements of $C^+(V)$ which remain fixed under J will be k if $n \leq 3$ and $C_0(V) = k + ke_M$ if $n = 4$. Notice that for $n = 4$ the algebra $C_0(V)$ is the center of $C^+(V)$. The elements of $C^-(V)$ which are fixed under J are in all cases just the elements of V.

If $\alpha \in C^+(V)$, then the norm $N\alpha = \alpha \circ \alpha^J$ is also in $C^+(V)$ and remains fixed under J since

$$(\alpha \circ \alpha^J)^J = \alpha^{J^2} \circ \alpha^J = \alpha \circ \alpha^J.$$

We conclude that $N\alpha \in k$ if $n \leq 3$ and $N\alpha \in C_0(V)$ if $n = 4$. For $\alpha, \beta \in C^+(V)$ we have, therefore, $N(\alpha \circ \beta) = \alpha \circ \beta \circ \beta^J \circ \alpha^J = N\beta \circ \alpha \circ \alpha^J = N\alpha \circ N\beta$.

We know that a regular element $\alpha \in C^+(V)$ must satisfy $N\alpha \in k^*$. Suppose conversely that $\alpha \in C^+(V)$ and $N\alpha = \alpha \circ \alpha^J = a \in k^*$. The element $(1/a)\, \alpha^J$ is then the inverse α^{-1} of α and we have for $X \in V$

$$X^\alpha = \frac{1}{a}(\alpha \circ X \circ \alpha^J).$$

X^α will lie in $C^-(V)$ and, since $(\alpha \circ X \circ \alpha^J)^J = \alpha \circ X \circ \alpha^J$, it is left fixed by J. Hence $X^\alpha \in V$. For $n \leq 3$, $N\alpha$ is always in k. Should α have an inverse, then $N\alpha \in k^*$ as $N\alpha \circ N\alpha^{-1} = 1$ shows. The regular elements of $C^+(V)$ are, therefore, the elements with a norm $\neq 0$ if $n \leq 3$. Should $n = 4$, one has to add the condition $N\alpha \in k^*$ since one only knows $N\alpha \in C_0(V)$. The point is that $X^\alpha \in V$ is automatically satisfied.

We had mapped the group $D(V)/k^*$ (Theorem 5.11) isomorphically onto $O^+(V)$ by mapping the coset αk^* onto the element $\sigma_\alpha \in O^+(V)$ given by $\sigma_\alpha(X) = X^\alpha$. For the spinorial norm $\theta(\sigma_\alpha)$ it follows by definition and formula (5.17) that $\theta(\sigma_\alpha) = N\alpha \cdot k^{*2}$. The spinorial norm corresponds, therefore, to the norm map in the group $D(V)/k^*$. Let us denote by $D_2(V)$ the group of those $\alpha \in C^+(V)$ for which $N\alpha \in k^{*2}$. Then we have the isomorphism

$$O'(V) \simeq D_2(V)/k^*.$$

Since we can change α in the coset αk^* by an element of k^* we can assume of α that $N\alpha = 1$. Such an α is determined up to a sign and we can also write

$$O'(V) \simeq D_0(V)/\{\pm 1\}$$

where $D_0(V)$ is the set of all elements of $C^+(V)$ with norm 1.

We discuss now the cases separately.

I. $n = 2$.

If A_1, A_2 is an orthogonal basis of V, then $e_M \circ e_M = -A_1^2A_2^2 = -G$ where G is the discriminant of V. We find $e_M^J = -e_M$. The algebra $C^+(V)$ coincides with $C_0(V) = k + ke_M$ and thus is commutative. For $\alpha = a + be_M$ we get $\alpha^J = a - be_M$ and

$$N\alpha = a^2 + Gb^2.$$

If $N\alpha \neq 0$, then α is regular and, consequently, $\alpha = A \circ B$ where A and B are non-isotropic vectors of V. Then $N\alpha = A^2B^2$, $A^{-1} = (1/A^2)A$, $B^{-1} = (1/B^2)B$ and

$$X^\alpha = \frac{1}{A^2B^2}(A \circ B \circ X \circ B \circ A).$$

As elements of $C^+(V)$ the products $B \circ X$ and $B \circ A$ commute and we may write

$$X^\alpha = \frac{1}{N\alpha}(A \circ B \circ A \circ B \circ X) = \frac{\alpha^2}{N\alpha} \circ X$$

where α^2 stands, of course, for $\alpha \circ \alpha$. Hence

$$X^\alpha \circ X = \frac{X^2}{N\alpha} \cdot \alpha^2$$

(X^2 is unambiguous and in k). If we apply J,

$$X \circ X^\alpha = \frac{X^2}{N\alpha} \cdot (\alpha^2)^J$$

hence

$$2(XX^\alpha) = (X \circ X^\alpha) + (X^\alpha \circ X) = \frac{X^2}{N\alpha} S(\alpha^2).$$

The symbol $S(\beta)$ means $\beta + \beta^J$ which is an element of k. The final result is

$$XX^\alpha = X^2 \cdot \frac{1}{2} S\left(\frac{\alpha^2}{N\alpha}\right).$$

The geometric meaning of this formula becomes clear if we consider the case of a euclidean plane when $k = R$, the field of real numbers, $G = 1$ and when $C^+(V)$ is isomorphic to the field of complex numbers. If θ is the angle of the complex number α, then $\alpha^2/N\alpha = e^{2i\theta}$ and $\frac{1}{2}S(\alpha^2/N\alpha) = \cos 2\theta$. Our formula means then that 2θ is the angle of the rotation $X \to X^\alpha$.

Leaving the trivial case of a hyperbolic plane to the reader we can generalize this result. If V is not a hyperbolic plane, then $-G$ is not a square and $C^+(V)$ is isomorphic to the quadratic extension field $k(\sqrt{-G}) = K$.

We have

$$O_2^+(V) \simeq K^*/k,$$

$$O_2'(V) = \Omega_2(V) \simeq D_0/\{\pm 1\},$$

where D_0 is the group of elements of K with norm 1. If $\sigma_\alpha \varepsilon O_2^+(V)$, then XX^α/X^2 does not depend on X, its value is $\frac{1}{2}S(\alpha^2/N\alpha)$, the analogue of the cosine of the angle of rotation.

II. $n = 3$.

If A_1, A_2, A_3 is an orthogonal basis of V, then $C^+(V)$ is generated by $1, i_1, i_2, i_3$ where $i_1 = A_2 \circ A_3$, $i_2 = A_3 \circ A_1$, $i_3 = A_1 \circ A_2$.

The table of multiplication is:

$$(i_1 \circ i_2) = -(i_2 \circ i_1) = -A_3^2 i_3, \qquad i_1^2 = -A_2^2 A_3^2,$$

$$(i_2 \circ i_3) = -(i_3 \circ i_2) = -A_1^2 i_1, \qquad i_2^2 = -A_3^2 A_1^2,$$

$$(i_3 \circ i_1) = -(i_1 \circ i_3) = -A_2^2 i_2, \qquad i_3^2 = -A_1^2 A_2^2.$$

An algebra of this form is called a generalized quaternion algebra.

The map J is very simple: 1 is kept fixed and each i_ν changes the sign. Thus, if $\alpha = x_0 + x_1 i_1 + x_2 i_2 + x_3 i_3$,

$$N\alpha = (x_0 + x_1 i_1 + x_2 i_2 + x_3 i_3)(x_0 - x_1 i_1 - x_2 i_2 - x_3 i_3)$$

$$= x_0^2 + A_2^2 A_3^2 x_1^2 + A_3^2 A_1^2 x_2^2 + A_1^2 A_2^2 x_3^2.$$

We merely repeat: $D(V)$ are the quaternions with norm $\neq 0$, $D_2(V)$ those whose norm is in k^{*2} and $D_0(V)$ those with norm 1. Then

$$O_3^+(V) \simeq D(V)/k^*, \quad \Omega_3(V) = O_3'(V) \simeq D_2(V)/k^* \simeq D_0(V)/\{\pm 1\}.$$

A special case is of interest. Suppose V contains isotropic vectors. Instead of a hyperbolic pair N, M we use $A_1 = N + \frac{1}{2}M$, $A_2 = N - \frac{1}{2}M$ and A_3, a vector orthogonal to A_1 and A_2. Since

$$A_1^2 = 1, \quad A_2^2 = -1, \quad A_3^2 = a \,\varepsilon\, k^*$$

we have

$$(i_1 \circ i_2) = -(i_2 \circ i_1) = -ai_3\,, \qquad i_1^2 = +a,$$
$$(i_2 \circ i_3) = -(i_3 \circ i_2) = -i_1\,, \qquad i_2^2 = -a,$$
$$(i_3 \circ i_1) = -(i_1 \circ i_3) = i_2\,, \qquad i_3^2 = +1.$$

It is easy to find 2-by-2 matrices which multiply according to the following rules:

$$1 = \begin{pmatrix} 1 & 0 \\ 0 & 1 \end{pmatrix}, \quad i_1 = \begin{pmatrix} 0 & a \\ 1 & 0 \end{pmatrix}, \quad i_2 = \begin{pmatrix} 0 & a \\ -1 & 0 \end{pmatrix}, \quad i_3 = \begin{pmatrix} 1 & 0 \\ 0 & -1 \end{pmatrix}.$$

They span the algebra of all 2-by-2 matrices over k and this algebra is, therefore, isomorphic to $C^+(V)$. The norm becomes a homomorphism of the matrices into k which is of degree 2 in the elements of the matrices and is merely the determinant. We see therefore

$$O_3^+(V) \simeq PGL_2(k), \qquad \Omega_3(V) \simeq PSL_2(k).$$

THEOREM 5.20. *If* $\dim V = 3$ *and if* V *contains isotropic vectors, then* $\Omega_3(V) \simeq PSL_2(k)$, *a simple group unless* $k = F_3$ *(the characteristic is* $\neq 2$).

III. $n = 4$.

We denote the center $C_0(V) = k + ke_M$ of $C^+(V)$ by K, the elements of K which have an inverse by K^* and their squares by K^{*2}. J leaves every element of K fixed and we have, consequently, $N\alpha = \alpha^2$ for $\alpha \in K$. An element $\alpha \in K$ ($\alpha = a + be_M$) will be regular if $\alpha^2 \in k^*$ and this implies $(a^2 + b^2e_M^2) + 2abe_M \in k^*$; either a or b must be 0, the coset $\alpha \cdot k^*$ is either k^* or $e_M \cdot k^*$. The corresponding rotation $\sigma_\alpha(X) = X^\alpha$ is either identity or the one induced by $e_M = A_1 \circ A_2 \circ A_3 \circ A_4$:

$$X^{e_M} = \tau_{A_1}\tau_{A_2}\tau_{A_3}\tau_{A_4}(X) = -X.$$

The regular elements of K correspond, therefore, to the rotations $\pm 1_V$.

Denote by P the group of all elements of $C^+(V)$ which have an inverse. We compose the map $O_4^+(V) \to D(V)/k^*$ with the map $D(V)/k^* \to P/K^*$ where the coset αk^* is mapped onto the coset αK^*. The coset αk^* will lie in the kernel if $\alpha \in K^*$; since α must be regular, this corresponds to the rotations $\pm 1_V$. We obtain the into map

$$O_4^+(V) \to P/K^* \qquad \text{whose kernel is } \pm 1_V$$

and, by restriction,

$$O_4'(V) \to P/K^* \qquad \text{whose kernel is the center of } O_4'(V).$$

The image αK^* of an element $\sigma_\alpha \varepsilon O_4^+(V)$ is a coset where $N(\alpha K^*) \subset k^* K^{*2}$. If, conversely, αK^* has this property, then $N\alpha = a\beta^2$ where $a \varepsilon k^*$, $\beta \varepsilon K^*$. The generator α of the coset αK^* may be changed by the factor β and we can achieve $N\alpha \varepsilon k^*$ which shows our coset to be the image of σ_α. Should $\sigma_\alpha \varepsilon O_4'(V)$, then $N\alpha \varepsilon k^{*2}$, hence $N(\alpha K^*) = K^{*2}$; for a coset with this property one can achieve $N\alpha = 1$ so that such a coset is indeed the image of some $\sigma_\alpha \varepsilon O_4'(V)$.

If we factor out the kernels of the maps we get isomorphisms into:

$$PO_4^+ \to P/K^* \quad \text{and} \quad PO_4' \to P/K^*$$

where the images of PO_4^+ are the cosets αK^* whose norms are in $k^* K^{*2}$ and those of PO_4' have norms in K^{*2}.

If $\sigma_\alpha \varepsilon O_4^+$, then σ_α and $-\sigma_\alpha$ have the same image αK^*. Their spinorial norms $\theta(\sigma_\alpha)$ and $\theta(-\sigma_\alpha)$ are determined by the norms of the coset $\alpha K^* : N\alpha \cdot K^{*2}$; indeed $\alpha K^* = \alpha e_M K^*$ and $N\alpha \cdot k^{*2}$, $N(\alpha e_M) \cdot k^{*2}$ are the spinorial norms of σ_α and $-\sigma_\alpha$.

Now we take a look at $C^+(V)$. We need only $i_1 = A_2 \circ A_3$, $i_2 = A_3 \circ A_1$, $i_3 = A_1 \circ A_2$ which generate the quaternion algebra of the three-dimensional subspace $W_0 = \langle A_1, A_2, A_3 \rangle$ of V. With this quaternion algebra $C^+(W_0)$ and $e_M = A_1 \circ A_2 \circ A_3 \circ A_4$, an element of the center $C_0(V)$, we can get the missing products containing A_4. Indeed, $(A_2 \circ A_3) \circ e_M = i_1 \circ e_M$ will be a scalar multiple of $A_1 \circ A_4$. This means now that

$$C^+(V) = C^+(W_0) + C^+(W_0) \circ e_M .$$

Another way of interpreting this fact is to say that $C^+(V)$ is the quaternion algebra of W_0 but with coefficients from $C_0(V)$ instead of k.

To get more information we must consider two possibilities.

a) G is not a square in k.

We remark that $e_M \circ e_M = A_1^2 A_2^2 A_3^2 A_4^2 = G$ so that e_M can be regarded as $\sqrt{G}$ and K as the quadratic extension field $k(\sqrt{G})$ of k. K^* consists now of the non-zero elements of K. If we call W the space obtained from W_0 by enlarging k to K, then the quaternion algebra $C^+(W)$ may be identified with $C^+(V)$. The elements of the group P are merely $D(W)$ and the elements whose norm is in K^{*2} form $D_2(W)$. Since $D(W)/K^*$ is isomorphic to $O_3^+(W)$ and $D_2(W)/K^*$ to $\Omega_3(W)$, we obtain an isomorphism into

$$PO_4^+(V) \to O_3^+(W).$$

The element σ_α of $O_4^+(V)$ and $-\sigma_\alpha$ are mapped onto the corresponding rotation $\lambda_\alpha \in O_3^+(W)$. The spinorial norms of σ_α and $-\sigma_\alpha$ are $\theta(\sigma_\alpha) = N\alpha \cdot k^{*2}$ respectively $\theta(-\sigma_\alpha) = \theta(\sigma_{\alpha e_M}) = N\alpha \cdot G \cdot k^{*2}$. The image λ_α is a rotation whose spinorial norm $\theta(\lambda_\alpha) = N\alpha \cdot K^{*2}$ lies in $k^* \cdot K^{*2}$ and $\theta(\lambda_\alpha) = \theta(\sigma_\alpha) \cdot K^{*2}$ gives the interrelation. As for $PO_4'(V)$, we obtain the isomorphism

$$O_4'(V) = PO_4'(V) \simeq \Omega_3(W).$$

Thus the orthogonal (projective) four-dimensional groups are isomorphic to certain three-dimensional ones. Unfortunately one does not get $P\Omega_4(V)$.

Again an important special case:

Let V contain isotropic vectors. V is certainly not hyperbolic since $G = 1$ is then a square. On the other hand, if G is a square and V contains the hyperbolic pair N, M, we could set $V = \langle N, M \rangle \perp U$. The plane U has discriminant $-G$ since that of $\langle N, M \rangle$ is -1. Let A, B be an orthogonal basis of U. We can solve the equation $(xA + yB)^2 = x^2A^2 + y^2B^2 = 0$ in a non-trivial way since $-B^2/A^2 = G/(A^2)^2$ is a square. This shows that V is hyperbolic if G is a square. Theorem 5.20 now implies

THEOREM 5.21. *Let* dim $V = 4$ *and the index of* V *be* 1. *This means that there are isotropic vectors and that G is not a square. Then*

$$O_4'(V) = \Omega_4(V) \simeq PSL_2(k(\sqrt{G})),$$

a simple group in all cases.

b) $G = c^2$ is a square in k.

We may change A_4 by a factor and achieve $G = 1$. Then $e_M^2 = 1$. Let $W_0 = \langle A_1, A_2, A_3 \rangle$ (just over k as field) be the subspace of V orthogonal to A_4 and $C^+(W_0)$ its quaternion algebra. Then $C^+(V) =$

$C^+(W_0) + C^+(W_0)e_M$. Put $u_1 = \frac{1}{2}(1 + e_M)$, $u_2 = \frac{1}{2}(1 - e_M)$, then we have $1 = u_1 + u_2$, $e_M = u_1 - u_2$, $u_1^2 = u_1$, $u_2^2 = u_2$, $u_1u_2 = 0$. We may write now (since $K = k + ke_M$)

$$C^+(V) = C^+(W_0)u_1 + C^+(W_0)u_2 , \qquad K = ku_1 + ku_2 .$$

Let $\alpha \in C^+(V)$, $\alpha = \beta u_1 + \gamma u_2$ with $\beta, \gamma \in C^+(W_0)$. The element α has an inverse if and only if β and γ have. The antiautomorphism J is identity on K and leaves, therefore, u_1 and u_2 fixed. We find

$$N\alpha = N\beta\cdot u_1 + N\gamma\cdot u_2 = \tfrac{1}{2}(N\beta + N\gamma) + \tfrac{1}{2}(N\beta - N\gamma)e_M .$$

$N\alpha \in k$ means $N\beta = N\gamma$, and then $N\alpha = N\beta = N\gamma$.

$$P = D(W_0)u_1 + D(W_0)u_2 , \qquad K^* = k^*u_1 + k^*u_2 ,$$

$$\alpha K^* = \beta k^* u_1 + \gamma k^* u_2 .$$

Hence the factor group P/K^* is described by the direct product of isomorphic groups,

$$D(W_0)/k^* \times D(W_0)/k^*,$$

and an element in P/K^* by a pair $(\beta k^*, \gamma k^*)$ of cosets, $\alpha K^* \to (\beta k^*, \gamma k^*)$.

Each $D(W_0)/k^* \simeq O_3^+(W_0)$, the coset βk^* being mapped onto an element $\lambda_\beta \in O_3^+(W_0)$ such that $\theta(\lambda_\beta) = N\beta\cdot k^{*2}$. We have, therefore, an isomorphism into:

$$PO_4^+(V) \to O_3^+(W_0) \times O_3^+(W_0)$$

and the images $(\lambda_\beta, \lambda_\gamma)$ of the element $\sigma_\alpha \in O_4^+(V)$ satisfy $\theta(\lambda_\beta) = \theta(\lambda_\gamma) = \theta(\sigma_\alpha)$. By restriction

$$PO_4'(V) \simeq \Omega_3(W_0) \times \Omega_3(W_0)$$

and $PO_4'(V)$ is thus a direct product.

We look again at the case in which V contains isotropic vectors; we may assume that W_0 does also contain isotropic vectors and know already that V is hyperbolic.

THEOREM 5.22. *If* $\dim V = 4$ *and if* V *is hyperbolic, then* $P\Omega_4(V) = PO_4'(V) \simeq PSL_2(k) \times PSL_2(k)$.

Another interesting case is that of a euclidean space of dimension 4 which we avoided in §2.

THEOREM 5.23. *If V is euclidean and* dim $V = 4$, *then* $PO_4^+(V) = P\Omega_4(V) = O_3^+(W_0) \times O_3^+(W_0)$, *a direct product of two simple groups since W_0 is euclidean.*

In relativity one studies a four-dimensional space V over the field R of real numbers with an orthogonal geometry based on the quadratic form $f = x_1^2 + x_2^2 + x_3^2 - x_4^2$. Since isotropic vectors exist, the group O_4^+/Ω_4 is isomorphic to R^*/R^{*2} which is of order 2. The discriminant is -1, not a square in R. Hence $P\Omega_4 = \Omega_4 \simeq PSL_2(R(i))$, the projective unimodular group of two dimensions over the field of complex numbers; Ω_4 is, therefore, a simple group which is called the Lorentz group. The reader may work out for himself that the Lorentz group consists of those rotations of V which move the vectors with $X^2 < 0$ but $x_4 > 0$ into vectors of the same type, in other words, rotations which do not interchange past and future.

7. *The structure of the group* $\Omega(V)$

In §6 we have elucidated the structure of the orthogonal groups for dimensions ≤ 4; we shall now investigate the higher dimensional cases. It will be convenient to adopt the following terminology: If U is a non-singular subspace of V and if $\tau \in O(U)$, then we shall identify τ with the element $\tau \perp 1_{U^*}$ of $O(V)$, merely write τ for this element (if there is no danger of a misunderstanding) and say that τ is an isometry of the subspace U. It is then clear what is meant by a plane or by a three-dimensional rotation of V.

Can we improve on Witt's theorem? We shall answer this question only for finite fields.

THEOREM 5.24. *Suppose that k is a finite field and τ an isometry of a subspace U_1 of V onto some subspace U_2 of V. If U_2 is contained in a non-singular subspace W of codimension ≥ 2, then we can extend τ to an element ρ of the group $\Omega(V)$.*

Proof: We can find a rotation σ of V which extends τ. If ρ is a rotation of W^*, then $\rho\sigma$ also extends τ. Since dim $W^* \geq 2$ and since k is a finite field, the space W^* contains vectors with arbitrarily given squares. One can find a ρ such that its spinorial norm $\theta(\rho)$ is equal to that of σ. Then $\theta(\rho\sigma) = 1$ which shows $\rho\sigma \in \Omega(V)$.

COROLLARY. If k is a finite field, dim $V \geq 4$ and N, M two non-zero isotropic vectors, then there exists an element $\lambda \in \Omega(V)$ such that $\lambda N = M$.

Proof: M is contained in some hyperbolic plane.

A similar but weaker statement is true for arbitrary fields:

THEOREM 5.25. *Let U_1 and U_2 be non-singular isometric subspaces of V and suppose that U_2 contains isotropic lines. There is a $\lambda \in \Omega(V)$ such that $U_2 = \lambda U_1$.*

Proof: We can find a rotation σ such that $U_2 = \sigma U_1$ and may follow it by any rotation ρ of U_2. Since U_2 contains isotropic lines, we can achieve $\theta(\rho) = \theta(\sigma)$. Setting $\lambda = \rho\sigma \in \Omega(V)$ we have $U_2 = \lambda U_1$.

COROLLARY. If $\dim V \geq 3$, and if N, M are isotropic vectors $\neq 0$, then there exists a $c \in k^*$ such that for any $d \in k^*$ an element $\lambda \in \Omega(V)$ can be found for which $\lambda N = cd^2M$.

Proof: Let M, M' be a hyperbolic pair. We can find a rotation σ such that $\sigma N = M$. Let $\theta(\sigma) = ck^{*2}$ and let ρ be the rotation of $\langle M, M'\rangle$ which sends M onto cd^2M. Then $\theta(\rho) = \theta(\sigma)$ and $\lambda = \rho\sigma$ is the desired element of $\Omega(V)$.

We use these corollaries to prove

LEMMA 5.4. *Let* $\dim V \geq 5$, $P = \langle A, B\rangle$ *a singular plane of V where $A^2 = B^2 \neq 0$. There exists a $\lambda \in \Omega(V)$ which keeps B fixed and sends A into a vector C such that $\langle A, C\rangle$ is a non-singular plane.*

Proof: If $\langle N\rangle$ is the radical of P, then $P = \langle B, N\rangle$ and we have $A = \alpha B + \beta N$; since $A^2 = B^2$ we get $\alpha = \pm 1$. We may replace αB by B and βN by N and obtain $A = B + N$.

Let $H = \langle B\rangle^*$ be the hyperplane orthogonal to B. The isotropic vector $N \neq 0$ belongs to H and we can find an isotropic vector M in H such that $NM = -A^2$. If k is finite, there exists a $\lambda \in \Omega(H)$ such that $\lambda N = M$ (we can extend λ to V). We obtain

$$\lambda B = B, \qquad \lambda A = B + M = C$$

and

$$AC = (B + N)(B + M) = B^2 + NM = A^2 - A^2 = 0.$$

Since $C^2 = A^2$, the plane $\langle A, C\rangle$ is non-singular. If k is infinite we can achieve $\lambda N = cd^2M$ for a suitable c and any d. We have $\lambda B = B$, $\lambda A = B + cd^2M = C$, $A^2 = C^2$, $AC = B^2 - cd^2A^2$. The plane

$\langle A, C\rangle$ is non-singular if $A^2C^2 - (AC)^2 \neq 0$ or $(A^2)^2 \neq (AC)^2$. We can choose d in such a way that

$$AC = B^2 - cd^2A^2 \neq \pm A^2.$$

We shall also need

LEMMA 5.5. *Let A be a non-isotropic vector of V and suppose σ is an isometry of V which keeps every line L generated by a vector B, satisfying $B^2 = A^2$, fixed. We can conclude $\sigma = \pm 1_V$, if k either contains more than 5 elements, or if $k = F_5$ but $n \geq 3$, or if $k = F_3$ but $n \geq 4$.*

Proof: Replacing σ by $-\sigma$, if necessary, we can achieve that $\sigma A = A$. Let $H = \langle A\rangle^*$ and assume first that H contains isotropic lines $\langle N\rangle$. Since $(A + N)^2 = A^2$, we have $\sigma(A + N) = \epsilon(A + N)$ with $\epsilon = \pm 1$. The image

$$\sigma N = \sigma((A + N) - A) = \epsilon(A + N) - A$$

of N must be isotropic which implies $\epsilon = +1$ and, therefore, $\sigma N = N\cdot$ An arbitrary vector X of H is contained in some hyperbolic plane of H and we see $\sigma X = X$, which, together with $\sigma A = A$, proves $\sigma = 1$.

If H is anisotropic, let $C \neq 0$ be a vector of H. If there are at least six rotations of the plane $P = \langle A, C\rangle$, they will move the line $\langle A\rangle$ into at least three distinct lines of P which, by our assumption, are kept fixed by σ. Recalling $\sigma A = A$, we find $\sigma C = C$. If k contains at least seven elements, this condition is satisfied for any vector C of H and we get again $\sigma = 1$. If $k = F_5$, then we may assume that H is anisotropic and, hence, dim $H = 2$. H contains two independent vectors C and D which satisfy $C^2 = D^2 = 2A^2$ and the planes $\langle A, C\rangle$ and $\langle A, D\rangle$ are anisotropic. Such planes have six rotations and we get again $\sigma = 1$. For $k = F_3$, we have supposed $n \geq 4$ and, hence, dim $H \geq 3$; H will contain isotropic vectors and our lemma is proved in all cases.

We are now in a position to prove

THEOREM 5.26. *Suppose* dim $V \geq 5$ *and let G be a subgroup of $O(V)$ which is invariant under transformation by elements of $\Omega(V)$, and which is not contained in the center of $O(V)$. Then the group G will contain an element $\sigma \neq 1$ which is the square of a three-dimensional rotation.*

Proof:

1) Select $\sigma \in G$, $\sigma \neq \pm 1_V$; σ must move some non-isotropic line $\langle A \rangle$. We first contend that, for a suitable σ, the plane $P = \langle A, \sigma A \rangle$ will be non-singular. Suppose, therefore, that P is singular. Since $(\sigma A)^2 = A^2$ we can use Lemma 5.4 and find a $\lambda \in \Omega(V)$ that keeps σA fixed and moves A into a vector $C = \lambda(A)$ for which the plane $\langle A, C \rangle$ is non-singular. The element $\rho = \lambda\sigma^{-1}\lambda^{-1}\sigma$ is also in G and we have

$$\rho(A) = \lambda\sigma^{-1}\lambda^{-1}(\sigma A) = \lambda\sigma^{-1}(\sigma A) = C.$$

2) We assume, therefore, that $P = \langle A, \sigma A \rangle$ is a non-singular plane. We shall construct an element $\rho \neq \pm 1_V$ of G which keeps some non-isotropic line fixed; we may assume that σ moves every non-isotropic line, otherwise we could take $\rho = \sigma$. There is a $\lambda \in \Omega(V)$ which keeps every vector of P fixed and does not commute with σ. Suppose we had found such a λ. Set $\rho = \lambda^{-1}\sigma^{-1}\lambda\sigma$; it is an element of G and

$$\rho(A) = \lambda^{-1}\sigma^{-1}\lambda(\sigma A) = \lambda^{-1}\sigma^{-1}(\sigma A) = \lambda^{-1}(A) = A.$$

Since $\rho(A) = A$ we have $\rho \neq -1_V$ and, since λ and σ do not commute, we have $\rho \neq 1_V$. To construct the desired λ we distinguish two cases. If $\sigma P \neq P$, let B be a vector of P such that $\sigma B \notin P$. We may write $\sigma B = C + D$, $C \in P$, $D \in P^*$ and $D \neq 0$. Since dim $P^* \geq 3$, there is a $\lambda \in \Omega(P^*)$ which moves the vector D. For this λ (extended to V) we find

$$\lambda\sigma(B) = \lambda(C + D) = C + \lambda D \neq C + D = \sigma B = \sigma\lambda(B);$$

hence $\lambda\sigma \neq \sigma\lambda$. If $\sigma P = P$, then $\sigma P^* = P^*$, but the restriction τ of σ to P^* is not in the center of $O(P^*)$ (since σ moves the non-isotropic lines). For a certain $\lambda \in \Omega(P^*)$ we have $\tau\lambda \neq \lambda\tau$, hence $\sigma\lambda \neq \lambda\sigma$.

3) Let, therefore, $\sigma \neq \pm 1_V$ be an element of G which keeps a certain non-isotropic line $\langle A \rangle$ fixed. By Lemma 5.5 there exists a vector B which satisfies $B^2 = A^2$ such that the line $\langle B \rangle$ is moved by σ. Set $\sigma B = C$, then $\langle B \rangle \neq \langle C \rangle$. We denote by τ_A the symmetry with respect to the hyperplane $\langle A \rangle^*$. By Witt's theorem there is a $\mu \in O(V)$ such that $\mu A = B$. Then $\mu\tau_A\mu^{-1} = \tau_B$, hence

$$\lambda = \tau_B\tau_A = \mu\tau_A\mu^{-1}\tau_A^{-1} \in \Omega(V).$$

We have $\sigma\tau_A\sigma^{-1} = \tau_{\sigma A} = \tau_A$ and $\sigma\tau_B\sigma^{-1} = \tau_C$, hence

$$\sigma\lambda\sigma^{-1} = \sigma\tau_B\sigma^{-1}\cdot\sigma\tau_A\sigma^{-1} = \tau_C\tau_A$$

and

$$\rho = \sigma\lambda\sigma^{-1}\lambda^{-1} = \tau_C\tau_A\tau_A^{-1}\tau_B^{-1} = \tau_C\tau_B .$$

This element ρ lies in G, it is $\neq 1$ since $\langle C\rangle \neq \langle B\rangle$, it belongs to $\Omega(V)$ since it is a commutator. The radical of the plane $P = \langle B, C\rangle$ has at most dimension 1 and is the radical of the space P^*; P^* contains, therefore, a non-singular subspace W of dimension $n - 3$. The element ρ keeps every vector of P^*, hence every vector of W, fixed. If U is the three-dimensional space orthogonal to W, then ρ is a rotation of U. It belongs to $\Omega(V)$, thus to $O'(V)$, and, therefore, to $O'(U) = \Omega(U)$. It follows that ρ is the square of a rotation of U. Our theorem is proved.

If V is anisotropic, then ρ is the square of a plane rotation and this exhausts our present knowledge of the structure of $\Omega(V)$ in this case.

If V contains isotropic lines one can say much more. Suppose the subspace U is anisotropic; then the axis of the rotation ρ is not isotropic, ρ is the square of a rotation of an anisotropic plane P. We shall prove that any anisotropic plane P can be imbedded in a three-dimensional non-singular subspace U' of V which contains isotropic lines; the element ρ may be regarded as a rotation of U' and we see that it is no loss of generality to assume that the original U contains isotropic lines. Select an isotropic vector $N \neq 0$ of V which is not orthogonal to P; this is possible since, otherwise, all isotropic lines of V would be in P^* and, therefore, left fixed by the rotations of P. We set $U' = P + \langle N\rangle$; if U' were singular and $\langle M\rangle$ its radical, then $\langle M\rangle$ would be the only isotropic line of U', $\langle M\rangle = \langle N\rangle$, contradicting the choice of N.

This shows also that the generators $(\tau_A\tau_B)^2$ of $\Omega(V)$ (see Theorem 3.23) are squares of rotations of three-dimensional subspaces U which contain isotropic lines.

Let us return to our element $\rho \neq 1$ of G and suppose the field k contains more than three elements. The group $\Omega(U)$ is simple (Theorem 5.20), $G \cap \Omega(U)$ contains ρ and is an invariant subgroup of $\Omega(U)$; this implies $\Omega(U) \subset G$. The subspace U contains a hyperbolic plane P and G contains $\Omega(P)$. Let U' be another three-dimensional subspace with isotropic lines; it contains a hyperbolic plane P'. By Theorem 5.25 we can find a $\lambda \varepsilon \Omega(V)$ such that $P' = \lambda P$. Since $\Omega(P') = \lambda\Omega(P)\lambda^{-1}$ we conclude that $\Omega(P') \subset G$. The group $\Omega(U')$ is simple and $\Omega(U') \cap G$ contains $\Omega(P')$. It follows that $\Omega(U') \subset G$. The group G contains all generators of $\Omega(V)$ and consequently all of $\Omega(V)$.

If $k = F_3$, we use the fact that a space over a finite field contains proper subspaces with any prescribed geometry. Since dim $V \geq 5$, we can find a subspace V_0' of dimension 4 which is of index 1 (is not a hyperbolic space). The subspace V_0' contains a three-dimensional subspace U' which is isometric to U. Using Witt's theorem we see that U is contained in a four-dimensional space V_0 which is of index 1. The group $\Omega(V_0)$ is simple (Theorem 5.21) and $\Omega(V_0) \cap G$ contains ρ. Hence $\Omega(V_0) \subset G$ and we can conclude again that $\Omega(V_1) \subset G$ where V_1 is any four-dimensional subspace which is of index 1. If U' is any three-dimensional subspace of V, we can imbed U' in such a space V_1 and deduce $\Omega(U') \subset G$. It follows that $\Omega(V) \subset G$ and we have proved

THEOREM 5.27. *Suppose* dim $V \geq 5$, V *a space with isotropic lines, and let* G *be a subgroup of* $O(V)$ *which is invariant under transformation by* $\Omega(V)$ *and which is not contained in the center of* $O(V)$. *Then* $\Omega(V) \subset G$. *If* $P\Omega(V)$ *is the factor group of* $\Omega(V)$ *modulo its center, then* $P\Omega(V)$ *is simple.*

This fundamental theorem was first proved by L. E. Dickson in special cases and then by J. Dieudonné in full generality. Let us look at some examples.

1) $k = R$, the field of real numbers. If dim $V = n \geq 3$ and if the geometry is based on the quadratic form $\sum x^2 - \sum y^2$, then the geometry is determined by n and by the number r of negative terms. The case $r = 0$ is disposed of by using the arguments in §2 and Theorem 5.23; the case $r = n$ leads to the same groups and is also settled. In every other case isotropic lines are present and Theorems 5.20, 5.21, 5.22, 5.27 show that $P\Omega(V)$ is simple unless V is a hyperbolic space of dimension 4. Since R^*/R^{*2} is of order 2, $\Omega(V)$ is of index 2 in $O^+(V)$ if $0 < r < n$. The discriminant of V is $(-1)^r$; the element -1_V belongs to $\Omega(V)$ if and only if both n and r are even.

2) $k = F_q$, a field with q elements. The group k^*/k^{*2} is again of order 2 so that $\Omega(V)$ is of index 2 in $O^+(V)$, even if V is an anisotropic plane (since the squares of the vectors $\neq 0$ of V range over k^*).

The order of $\Omega(V)$ is, therefore, half the order of $O^+(V)$. To get the order of $P\Omega(V)$ one would have to divide by 2 again whenever the discriminant of V is a square and n is even. For an even n we had attached a sign $\epsilon = \pm 1$ to the two possible geometries. If $\epsilon = +1$, then the discriminant is $(-1)^{n/2}$; if $\epsilon = -1$, then it is $(-1)^{n/2}g$ where g is a non-square of F_q. The reader can work out for himself that the discriminant is a square if and only if $q^{n/2} - \epsilon$ is divisible by 4. The order of $P\Omega_n(U)$ is, therefore, given by

$$\frac{1}{2}\, q^{(n-1)^2/4} \cdot \prod_{i=1}^{(n-1)/2} (q^{2i} - 1) \qquad \text{for } n \text{ odd},$$

$$\frac{1}{d}\, q^{n(n-2)/4}(q^{n/2} - \epsilon) \cdot \prod_{i=1}^{(n-2)/2} (q^{2i} - 1) \qquad \text{for } n \text{ even},$$

where d is the greatest common divisor of 4 and $q^{n/2} - \epsilon$.

In §1 we determined the order of $PSp_n(F_q)$; if q is odd, then the d appearing in this order is 2. Thus the orders of $PSp_{2m}(F_q)$ and of $P\Omega_{2m+1}(F_q)$ are equal. It has been shown that $PSp_4(F_q)$ and $P\Omega_5(F_q)$ are isomorphic groups but that the groups $PSp_{2m}(F_q)$ and $P\Omega_{2m+1}(F_q)$ are not isomorphic if $m \geq 3$. One has, therefore, infinitely many pairs of simple, non-isomorphic finite groups with the same order. These pairs and the pair with the order 20160 which we have discussed in Chapter IV are the only pairs known to have this property; it would, of course, be an extremely hard problem of group theory to prove that there are really no others.

Should V be anisotropic, then the structure of $\Omega(V)$ is known only for special fields k. M. Kneser and J. Dieudonné have proved that $P\Omega(V)$ is always simple if dim $V \geq 5$ and if k is an algebraic number field. This is due to the fact that V can not be elliptic for a valuation of k if k is a number field and if dim $V \geq 5$. It is of course conceivable that for arbitrary fields a similar result can be proved. One might have to assume that V is not elliptic for any of the valuations of k and possibly dim $V \geq 5$.

Exercise. Let us call a group G strongly simple if one can find an integer N with the following property: Given any element $\sigma \neq 1$ of G; every element λ of G is a product of at most N transforms $\tau\sigma^{\pm 1}\tau^{-1}$ of $\sigma^{\pm 1}(\tau \in G)$. Show that, apart from trivial cases, the groups $PSL_n(k)$, $PSp_n(k)$ and $P\Omega_n$ (if V contains isotropic lines) are strongly simple provided that k is commutative. Show that $O_3^+(V)$ is not strongly simple if V is a euclidean space. It is doubtful whether $PSL_n(k)$ is strongly simple for all non-commutative fields. What is the trouble? A reader acquainted with topological groups may prove that an infinite compact group can not be strongly simple.

BIBLIOGRAPHY AND SUGGESTIONS FOR FURTHER READING

We shall restrict ourselves to a rather small collection of books and articles. A more complete bibliography can be found in [5] which is the most important book to consult.

[1] Baer, R., *Linear Algebra and Projective Geometry*, Academic Press, New York, 1952.

[2] Chevalley, C., *The Algebraic Theory of Spinors*, Columbia University Press, New York, 1954.

[3] Chow, W. L., *On the Geometry of Algebraic Homogeneous Spaces*, Ann. of Math., Vol. 50, 1949, pp. 32–67.

[4] Dieudonné, J., *Sur les groupes classiques*, Actualités Sci. Ind., No. 1040, Hermann, Paris, 1948.

[5] Dieudonné, J., *La géométrie des groupes classiques*, Ergebnisse der Mathematik und ihrer Grenzgebiete, Neue Folge, Heft 5, Springer, Berlin, 1955.

[6] Eichler, M., *Quadratische Formen und orthogonale Gruppen*, Springer, Berlin, 1952.

[7] Lefschetz, S., *Algebraic Topology*, Colloquium Publ. of the Amer. Math. Society, 1942.

[8] Pontrjagin, L., *Topological Groups*, Princeton Math. Series, Princeton University Press, Princeton, 1939.

[9] van der Waerden, B. L., *Moderne Algebra*, Springer, Berlin, 1937.

[10] Contributions to Geometry, The Fourth H. E. Slaught Mem. Paper, Supplement to the Amer. Math. Monthly, 1955.

To Chapter I: A deficiency in elementary algebra can be remedied by a little reading in any book on modern algebra, for instance [9]. The duality theory of vector spaces as well as of abelian groups (§4 and §6) can be generalized if one introduces topological notions. See for instance [7], Chapter II, and [8].

To Chapter II: For projective geometry [1] may be consulted. I suggest, for the most fascinating developments in projective geometry, a study of [3]; read [5], Chapter III, for a survey. We have completely left aside the interesting problem of non-Desarguian planes. The reader will get an idea of these problems if he looks at some of the articles in [9]. He will find there some bibliography.

To Chapters III–V: We have mentioned the important case of a unitary geometry and have only casually introduced orthogonal geometry if the characteristic is 2. The best way to get acquainted with these topics is again to read [5]; in [2] and [6] more details on orthogonal groups will be found.

INDEX[1]

[1]The Index contains the main definitions.